ANTIMICROBIAL RESISTANCE

Collaborative Measures of Control

Edited by

Sunil D. Saroj

CRC Press
Taylor & Francis Group
Boca Raton London New York

CRC Press is an imprint of the
Taylor & Francis Group, an **informa** business

First edition published 2023
by CRC Press
6000 Broken Sound Parkway NW, Suite 300, Boca Raton, FL 33487-2742

and by CRC Press
4 Park Square, Milton Park, Abingdon, Oxon, OX14 4RN

CRC Press is an imprint of Taylor & Francis Group, LLC

ISBN: 9781032321615 (hbk)
ISBN: 9781032321653 (pbk)
ISBN: 9781003313175 (ebk)

DOI: 10.1201/9781003313175

Publisher's note: This book has been prepared from camera-ready copy provided by the authors.

Preface

At present, we are running out of antibiotics, and despite the presence of current antibiotics, bacterial infectious diseases are the leading causes of death worldwide. Antibiotics have served humanity by saving human life from severe bacterial infections. Also, antibiotics play a pivotal role in medical and surgical processes involving complex immune-suppressive regimes, particularly in chemotherapy for cancer treatments and organ transplants. Bacteria have evolved several antimicrobial resistance mechanisms through which they can either chemically modify the antibiotics or pump the antibiotics out of their system, rendering the therapeutic procedures ineffective. There is a global increase in drug-tolerant bacteria, which, combined with the absence of novel antimicrobial targets, has raised a sense of emergency in public health management, reasoned to the increase in life-threatening situations. In addition to the rise in the number of deaths, the problem of antibiotic resistance has imposed a significant economic burden on the healthcare system and patients. The rate of identification of novel antibiotics is much slower than the rate of evolution of antimicrobial resistance. Therefore, there is a need for alternate novel strategies to combat antimicrobial resistance. One of the strategies that could bring about pivotal change is behavioural change. Thus, in this book, we emphasize that in addition to the obligation to develop novel antimicrobials, there is an immediate need for awareness about antimicrobial resistance among the population. The topics on global initiatives to combat AMR will assist the readers in understanding the concerted approach taken by the world communities. We further discuss the effectiveness of education in behavioural change to fight AMR. The book also details the difficulties in implementing global initiatives to prevent AMR.

Acknowledgment

PREVENTiT - Prevention and Risk Management of Antimicrobial Resistance is funded by the ERASMUS+ program of the European Union (598515-EPP-1-2018-1-IN-EPPKA2-CBHE-JP). The consortium comprises experts from Amrita University, Chitkara University, Kalinga Institute of Industrial Technology, Maastricht University, Manipal Academy of Higher Education, Rural Institute for People's Enlightenment, Rīga Stradiņš University, Symbiosis International (Deemed University), Universidade Católica Portuguesa, University of Milan, Voluntary Health Association of Tripura.

Co-funded by the Erasmus+ Programme of the European Union

Disclaimer

This project (Project reference: 598515-EPP-1-2018-1-IN-EPPKA2-CBHE-JP) has been funded with support from the European Commission. This book reflects the views only of the author, and the Commission cannot be held responsible for any use which may be made of the information contained therein.

Contributors

Agnita Roychowdhury Symbiosis School of Biological Sciences, Symbiosis International (Deemed University) Pune, India

Amrita Bhagwat Symbiosis School of Biological Sciences, Symbiosis International (Deemed University) Pune, India

Bunty Sharma Department of Biotechnology, Maharishi Markandeshwar (Deemed to be) University, Mullana, Ambala

Deena Mariyam Kempegowda Institute of Medical Sciences, Bengaluru, India

Dona Boban Amrita Institute of Medical Science (AIMS), Kochi, India

Helmut Brand Department of International Health, Care and Public Health Research Institute – CAPHRI, Maastricht University, Maastricht, The Netherlands

Honey Tandon Global Health Student, Kings College, London, UK; Practicing Dietician (PGDDPHN), India, ex HOD (Dietetics) Max Hospitals and Columbia Asia, Gurugram, India

Himanshu Sekhar Pradhan KIIT School of Public Health, KIIT Deemed to be University, Bhubaneswar, Odisha, India

Jyoti Prakash KIIT School of Public Health, KIIT Deemed to be University, Bhubaneswar, Odisha, India

Kiranjeet Kaur Chitkara School of Health Sciences, Chitkara University, Punjab, India

Keerti Bhusan Pradhan Chitkara School of Health Sciences, Chitkara University, Punjab, India

Lauma Springe Department of International Business and Economics, Riga Stradins University, Latvia

Meenakshi Sood Chitkara School of Health Sciences, Chitkara University, Punjab, India

Mousumi Samal KIIT School of Public Health, KIIT Deemed to be University, Bhubaneswar, Odisha, India

Navneet Kaur Bains Chitkara School of Health Sciences, Chitkara University, Punjab, India

Prakash Narayanan Prasanna School of Public Health, Manipal Academy of Higher Education, Manipal, Karnataka, India

Preethi John Chitkara Global Health Institute, Chitkara University, Punjab, India

Rajashri Banerji Symbiosis School of Biological Sciences, Symbiosis International (Deemed University) Pune, India

Ranjika Bhattacharya Symbiosis School of Biological Sciences, Symbiosis International (Deemed University) Pune, India

Rakshitha K. Prasanna School of Public Health, Manipal Academy of Higher Education, Manipal, Karnataka, India

Riya Joshi Symbiosis School of Biological Sciences, Symbiosis International (Deemed University) Pune, India

Romans Putans Department of International Business and Economics, Riga Stradins University, Latvia

Sandhya Vashisht Project manager, Tech Mahindra SMART for healthcare, Mohali, India

Sanjay Pattanshetty Department of Health Policy, Prasanna School of Public Health, Manipal Academy of Higher Education, Manipal, Karnataka, India

Sanjeev K. Singh Amrita Institute of Medical Science (AIMS), Kochi, India

Stefano Greco Chitkara Spaak Centre for Multidisciplinary European Studies, Chitkara University Punjab, India

Sudhir Kumar Satpathy KIIT School of Public Health, KIIT Deemed to be University, Bhubaneswar, Odisha, India

Sunil D. Saroj Symbiosis School of Biological Sciences, Symbiosis International (Deemed University) Pune, India

Tiyasa Haldar Symbiosis School of Biological Sciences, Symbiosis International (Deemed University) Pune, India

Ujjayni Saha Symbiosis School of Biological Sciences, Symbiosis International (Deemed University) Pune, India

Vinay Kumari Chitkara University College of Nursing, Chitkara University, Himachal Pradesh, India

Contents

List of Figures

List of Tables

1

Microbial Threats - The AMR Pandemic

Rajashri Banerji

Symbiosis School of Biological Sciences, Symbiosis International (Deemed University) Pune, 412115, India

Sunil D. Saroj

Symbiosis School of Biological Sciences, Symbiosis International (Deemed University) Pune, 412115, India

CONTENTS

1.1 Introduction

Increasing antimicrobial resistance (AMR) is considered a global public health threat. Mostly surpassed by the occurrence of pandemics, health experts suggest that pandemic-induced antimicrobials used for the treatment of the disease are accelerating the AMR crisis [1]. AMR jeopardises healthcare advances, food production, and life expectancy. Preventing infections in the first place, decreasing the development of resistance through better antibiotic usage, and stopping the spread of resistance after it has developed, are all the necessary steps to be taken in combating the AMR crisis [2]. Although the world has collectively prevented some of the worst pandemics in history, (the Spanish Flu in 1918, killing nearly 40 million people worldwide), the emergence and re-emergence of a host of viruses in the last two decades are symptoms of new threats that confront the global health community [3]. AMR is now one of the most pressing public health challenges, and it threatens our ability to battle a variety of multidrug-resistant diseases. Without efficient antimicrobials, even the most common and usually curable infections can become difficult to treat, leading to life-threatening infections. As a result, various medical operations whose feasibility and success are heavily reliant on the availability of appropriate antimicrobials to prevent and treat infections may become more dangerous [4].

Although precise figures to justify the economic damage of each pandemic outbreak are challenging to come by, approximate estimates are available. According to estimations made in 2015, by 2050, AMR is anticipated to cause 10 million annual fatalities and over 100 trillion dollars in economic damages [5]. Thus, there is an urgent need to regulate such a hazardous calamity to occur. Several policies across the globe have been implicated in controlling the overuse of antibiotics; however, awareness within the communities is still lacking. In this chapter, we discuss the role of pandemics in the emergence of microbial threats and the urgent need to regulate the rise in AMR before we are hit with another major pandemic.

1.2 Microbial threats

The rise in AMR is a global crisis that can threaten the use of antibiotics. Clean water is scarce in developing countries, while the urban metropolis expands exponentially. As a result, infectious diseases disproportionately impact poorer countries in the developing world, becoming a worldwide burden. Infectious diseases are the major cause of death for children and young adults in underdeveloped nations, accounting for one out of every two deaths. Several pathogens have been identified which are considered as potential microbial threats (Table 1.1). In 2015, World Health Organisation (WHO) had identified 10 major emerging viruses which required primary attention (Table 1.2) [6]. While in 2017, the WHO listed priority pathogens that required immediate attention for the development of antibiotics based on the priority (Table 1.3) [7]. In 2019, Centres for Disease Control and Prevention (CDC) declared 18 microbial threats and categorised them as urgent, serious, and concerning (Table 1.4) [2].

As a result of the inability to treat disease due to emergence in microbial threats, there will be an increase in mortality and morbidity. This will increase the risk of drug-resistant infections worldwide, healthcare expenses, more prolonged and more frequent hospital admissions, and alternative medicine research and development costs.

1.3 Key drivers of emerging microbial threats

Microbial threats can emerge, remerge or persist. Few microbial threats may cause previously unidentified disease, while other known pathogens might infect a larger population, spreading around different geographical areas. A varying set of factors regulates such emerging microbial threats. Moreover, lower/ middle-income countries are at a higher risk for AMR due to poorly resourced health programs, healthcare services, and inefficacy in controlling the consumption of antibiotics due to self-medication [8].

TABLE 1.1
Potential microbial threats

Microbial threat	Infections
Category A	
Bacillus anthracis	Anthrax
Clostridium botulinum	Botulism
Yersinia pestis	Plague
	Small pox
Francisella tularensis	Tularemia
Filoviruses	Viral hemorrhagic fever
Category B	
Brucella	Brucellosis
Clostridium perfringens	Infections of skin and tissues
Salmonella	Gastrointestinal tract infections
Escherichia coli	Gastrointestinal tract infections
Shigella	Shigellosis
Burkholderia pseudomallei	Melioidosis
Chlamydia psittaci	Psittacosis
Coxiella burneti	Q fever
Staphylococcus aureus	Pneumonia, infections of wound, bloodstream, urinary tract
Rickettsia prowazekii	Typhus fever
Alphaviruses	Viral encephalitis
Category C	
Nipah virus	Acute respiratory illness, encephalitis
Hentavirus	Respiratory tract infection
Influenza	Respiratory tract infection

TABLE 1.2
Viral threats according to WHO

Microbial threat	Route of transmission	Infections
Nairovirus	Ticks	Crimean-Congo haemorrhagic fever
Filovirus (Ebola and Marburg)	Direct contact	haemorrhagic fever
Co-V and SARS	Direct contact	Respiratory tract infection
Lassa virus	Rat	Lassa fever
Nipah	Bat/ pigs	Respiratory tract infection, encephalitis
Rift valley fever virus	Cattle, mosquitoes	Rift valley fever
Chikungunya virus	Mosquito	Chikungunya virus
Phlebovirus	Ticks	Thrombocytopenia fever
Zika virus	Mosquito	Fever, rash, joint pain

A few of the factors that might lead to the emergence of microbial threats have been discussed below.

1.3.1 Genetic and environmental factors

Genetic and environmental factors allow microbial threats to adapt to stress and evolve. Such evolution processes under selective pressure enable microorganisms to persist and evade the host immune system [9]. The presence of mobile genetic elements (plasmids, transposons, integrons) contribute dissemination of AMR among several bacterial genera [10]. In addition to mutation, resistance also results from degradation or alteration of the antimicrobial, reduced uptake of an antimicrobial, active efflux of an antimicrobial out of the cell, and biofilm assisted AMR (Figure 1.1). Biofilms consist of polysaccharides, proteins, extracellular DNA and lipids, and their composition is dependent on strain, growth conditions and the age of the biofilm. The biofilm-specific environment accelerates the development of tolerant micro-organisms, which are distinct from those of planktonic cells [11]. During biofilm development in the presence of antibiotics, a transitory phenotype emerges as a result of diverse responses within a genetically identical bac-

TABLE 1.3
Priority pathogens according to WHO

Microbial threat	Infection
Critical priority	
Carbapenem-resistant *Acinetobacter baumannii*	Pneumonia, infections of wound, bloodstream, urinary tract
Carbapenem-resistant *Pseudomonas aeruginosa*	Infection of skin and throat
Carbapenem-resistant ESBL-producing *Enterobacteriaceae*	Hospital acquired infections
High priority	
Vancomycin-resistant *Enterococcus faecium*	Hospital acquired infections
Methicillin-resistant, vancomycin-intermediate/resistant *Staphylococcus aureus*	Pneumonia, infections of wound, bloodstream, urinary tract
Clarithromycin-resistant *Helicobacter pylori*	Gastrointestinal tract infections
Fluoroquinolone-resistant *Campylobacter* spp	Diarrhoea, fever, abdominal cramps
Fluoroquinolone-resistant *Salmonella*	Diarrhoea, fever, abdominal cramps
Cephalosporin-resistant, Fluoroquinolone-resistant *Neisseria gonorrhoeae*	Vaginal infections
Medium priority	
Penicillin-non-susceptible *Streptococcus pneumoniae*	Pneumonia, infections of wound, bloodstream, urinary tract
Ampicillin-resistant *Haemophilus influenzae*	Respiratory tract infections
Fluoroquinolone-resistant *Shigella* spp	Diarrhoea, fever, abdominal cramps

TABLE 1.4
Categorisation of microbial threats according to CDC

Microbial threat	Infections
Urgent threats	
Carbapenem-resistant *Acinetobacter*	Pneumonia, infections of wound, bloodstream, urinary tract
Candida auris	Hospital acquired infections
Clostridioides difficile	Diarrhoea and colitis
Carbapenem-resistant Enterobacterales	Hospital acquired infections
Drug-resistant *Neisseria gonorrhoeae*	Vaginal infections
Serious threats	
Drug-resistant *Campylobacter*	Diarrhoea, fever, abdominal cramps
Drug-resistant *Candida*	Oral and vaginal infection
ESBL-producing Enterobacterales	Hospital acquired infections
Vancmycin-resistant Enterococci	Hospital acquired infections
Multidrug-resistant *Pseudomonas aeruginosa*	Infection of skin and throat
Drug-resistant nontyphoidal *Salmonella*	Diarrhoea, fever, abdominal cramps
Drug-resistant *Salmonella* Typhi	Typhoid fever
Drug-resistant *Shigella*	Diarrhoea, fever, abdominal cramps
Methicillin-resistant *Staphylococcus aureus*	Hospital acquired infections
Drug-resistant *Streptococcus pneumoniae*	Pneumoniae, infection of ear and sinus
Drug-resistant Tuberculosis	Tuberculosis
Concerning Threats	
Erythromycin-resistant Group A *Streptococcus*	Infection of skin and respiratory tract
Clindamycin-resistant Group B *Streptococcus*	Infection of reproductive tract

terial population. These are phenotypic varieties of bacteria that are not genetically resistant to antibiotics but are tolerant of high antibiotic concentrations known as persister cells. Antibiotics can kill the majority of bacterial cells. Still, persisters can survive and repopulate biofilms since they have a dormant condition that prevents the antibiotic targets from synthesising [12].

1.3.2 Poverty and population density

About 85 percent of the world's population live in low/ middle-income countries which are densely populated. A well-structured healthcare system is lacking in the low/ middle-income countries with high population density. All countries are affected by AMR; however, the impact is disproportionately heavier in low/ middle-income countries. Thus, controlling the usage of antibiotics and keeping no track of the antibiotic consumption increases the risk of AMR [4]. Moreover, within an overcrowded setting, the chances of AMR transmission through direct contact, bacterial shedding, or food are high [13]. A study correlated the income status of the countries to AMR prevalence and observed the presence of cephalosporin-resistant *Escherichia coli* and *Klebsiella* in the community [14]. In low/ middle-income nations, the challenge to ensure universal health care while protecting already accessible antimicrobials is a key concern. Also, in many low/ middle-income nations, the access to skilled healthcare providers is severely limited [15].

1.3.3 Inadequacy in infection prevention and control strategies

The lack of appropriate regulation and guidelines on the sales and distribution of antimicrobials and the consumption of antimicrobials without prescription can give rise to AMR. Similarly, unlicensed pharmacies are attracted more to the public due to their accessibility [16]. Moreover, poor storage of drugs can also affect the data that can accurately predict the antimicrobial dosage since degraded antimicrobial will provide less dynamic content [17]. Moreover, the doctors are overworked due to the high patient-to-doctor ratio. Thus, there is often limited time for proper communication with patients about drug adherence standards and the implications

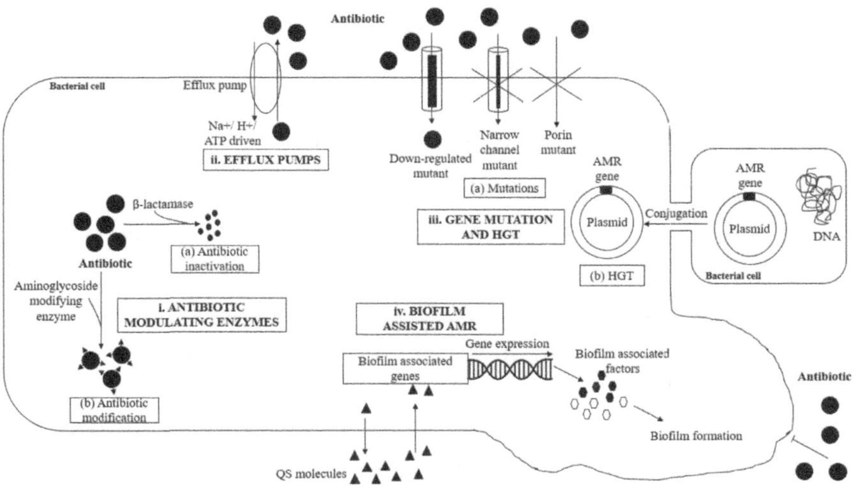

FIGURE 1.1
Mechanism of AMR resistance: (i-a) AMR can be achieved by modulating the antibiotic by inactivation or degradation of the antibiotic with the help of beta-lactamase. (i-b) Aminoglycoside modifying enzymes add a chemical group to the exposed sites of the antibiotic. (ii) Efflux pumps present on the membrane eject the antibiotic by transporting Na+/H+. Many efflux pumps require ATP for pumping out the antibiotic. (iii) Gene mutations and HGT also lead to AMR. (a) in transmembrane protein Porin affects the uptake of hydrophilic antibiotics where the mutation can down-regulate the function of Porin, narrow channel mutation, or total porin loss inhibits the uptake of the antibiotic. (b) With the help of conjugation (HGT), AMR gene encoding plasmid can be transferred to the bacteria, thus providing resistance against antibiotics. (iv) The high cell density bacteria secrete QS molecules leads to the auto-induction process that activates biofilm-associated genes. These factors lead to the formation of biofilm, which is resistant to antibiotics.

of poor or non-adherence to these guidelines. Without a confirmed diagnosis and reason for antimicrobial treatment, broad-spectrum antibiotics are sometimes used to treat patients [16]. Hence, medical personnel and scientists cannot be solely responsible for AMR control. However, continuous education is necessary for healthcare providers to understand the value of rational prescribing practice.

1.3.4 Lack of technological advances

Poor hygiene and sanitation due to lack of technological advances in low/ middle-income countries are often problematic, compromising healthcare. Moreover, it has been stated that industrial pollution and inefficacy in managing urban waste play an essential role in AMR rise in low/ middle-income countries [18]. Effective infection prevention programmes are also challenging to implement in national contexts with a lack of infection prevention training, personal protective equipment, and reliance on reprocessed single-use medical equipment. As a result of these resource constraints, rates of device-associated infections in low/ middle-income nations are higher [4]. Since traditional culturing methods include culturing specific organisms within laboratory conditions, this limits the number of studies to a particular group. Currently, advanced culture techniques known as culturomics have been used for high throughput analysis to identify and characterise microorganisms in their habitats, including unknown species that were previously unculturable [19]. Culturomics uses various conditions and techniques such as MALDI-TOF and 16S rRNA sequencing to identify species rapidly. However, with recent technological advancements and lower costs in next-generation sequencing technologies, culture-independent metagenomic research is more feasible [13]. However, unknown AMR pathogens go undetected due to a lack of modern tools, particularly in low/ middle-income countries. Health workers in rural places who do not have AMR testing have difficulty determining which antimicrobial to employ without an antibiotic sensitivity test. As a result, doctors increasingly prescribe broad-spectrum antibiotics to treat illnesses caused by many bacteria species or infections whose aetiology is difficult or time-consuming to determine [16].

As a result, public awareness campaigns and government measures to educate the community regarding the side effects and risks connected with antibiotic misuse are urgently needed.

1.4 Consequences of delayed regulation of microbial threats

Several pharmaceutical companies are continuously developing new antimicrobials; however, obtaining approval from the regulatory authority is often difficult. Bureaucracy, a lack of clarity, variances in clinical trial criteria between nations, changes in regulatory and licencing standards, and inadequate communication channels have all been identified as challenges in obtaining regulatory approval [20]. The regulatory authorities consider studies comparing antibiotics to placebo are as unethical. As a result, trials show that new treatments are non-inferior to existing drugs within a wide statistical margin. This necessitates a large sample population and, as a result, hefty expenditures, rendering antibiotic development uneconomical [21]. Hence, such regulatory obligations can lead to delays in launching new antimicrobials, which can further delay the treatments and increase the rate of mortality and morbidity.

1.5 Pandemic and its association with AMR

Pandemic leading to utilisation and consumption of antibiotics, which eventually contributes to rise in AMR. The two significant pandemics and their association with AMR for which no complete cure have been identified yet have been discussed below.

1.5.1 AIDS and AMR (1981-present)

HIV (Human Immunodeficiency Virus) causes AIDS (Acquired Immunodeficiency Syndrome), which affects the human host's immune system. Approximately 38 million people are infected with HIV worldwide, with the highest infection rate in sub-Saharan Africa [2]. When antiretroviral medicines were first offered for HIV treatment in 1987, HIV strains developed resistance. Since then, a slew of drug-resistant strains have emerged, each with a unique vulnerability to one of three types of antiretrovirals: nucleoside reverse-transcriptase inhibitors, non-nucleoside reverse-transcriptase inhibitors, or protease inhibitors [22]. People infected with HIV have an increased risk of contracting a bacterial infection. A meta-review analysis of studies published in 1995-2020 predicted the increased risk of AMR with MRSA (Methicillin-Resistant *Staphylococcus aureus*) strains in HIV infected patients. Similarly, increased resistance in *E. coli* and *K. pneumoniae* to cephalosporin was identified [23]. Moreover, as a preventive measure, preexposure prophylaxis (PrEP) is being used by HIV-negative people. However, it has been noted that the risk of bacterial sexually transmitted infections increases with PrEP [24]. Similarly, a population-level mathematical model revealed that oral preexposure prophylaxis and vaginal microbicides reduce new HIV infections. However, it increases resistance if administered to HIV-infected patients [25]. Another study reported the use of abacavir and azidothymidine for HIV, which reduced plasmid transmission in E. coli and *K. pneumoniae*, suggesting the use of these drugs to decolonise AMR plasmids in humans, animals, or the environment [26]. Alternatively, a study looked at the AMR patterns of clinical isolates of *Neisseria gonorrhoeae* in women who had HIV or were at high risk of getting infected. The isolates of *N. gonorrhoeae* were shown to be resistant to ciprofloxacin, tetracycline, ceftriaxone and cefixime. Furthermore, most of the women with *N. gonorrhoea* were asymptomatic, suggesting that they could be a reservoir for continuous transmission. To guarantee effective treatment and minimise antibiotic resistance progression, AMR testing of *N. gonorrhoeae* isolates is required [27]. Similarly, a study estimated the prevalence of ESBL (extended-spectrum beta-lactamase) *Enterobacteriaceae* and carbapenemase-producing *Enterobacteriaceae* in stool samples in HIV-infected children in Zimbabwe [28].

Even though significant advancements in combating AIDS have occurred in the past decade, antibiotic overuse is still a critical concern that deserves immediate attention.

1.5.2 COVID-19 and AMR (2019- present)

The coronavirus causing the coronavirus disease (COVID-19) continues even today, posing an immense threat worldwide infecting with over 240 million people resulting in over 4.8 million fatal cases [29]. COVID-19 has also been associated with bacterial and fungal infections, which has led to inappropriate usage of antibiotics by the human community [30].

Chryseobacterium indologenes, a gram-negative bacteria, is a nosocomial pathogen associated with immunocompromised patients and exhibits resistance towards penicillin, carbapenem and cephalosporin. In the past year, *C. indologenes* has been associated with COVID-19 infection causing bacteremia. The whole-genome sequencing of *C. indologenes* identified the existence of blaIND-2 (resistance towards carbapenems), blaCIA (resistance towards beta-lactamase) and blaCcrA, leading to the development of AMR [31]. A study in the Arab countries identified the misuse of azithromycin and antivirals such as lopinavir favipiravir during the COVID-19 treatment, which can give rise to AMR [32]. Other studies from Arab countries revealed the presence of carbapenem-resistant *Stenotrophomonas maltophilia*, *K. pneumoniae*, and *Pseudomonas aeruginosa* from COVID-19 patient isolates in Qatar [33]. While, Egypt showed carbapenem-resistant *Acinetobacter baumannii*, Enterococci, and Pseudomonas spp in COVID-19 patients [34]. Japan reported increased third-generation cephalosporin-resistant *K. pneumoniae* from COVID-19 patients [29], while India noted the presence of MRSA in blood and urine samples of COVID-19 patients [35]. A study in Sweden characterised the AMR for secondary bacterial infections/ co-infections during COVID-19 and reported the presence of multidrug-resistant *A. baumannii* as a significant pathogen among the COVID-19 patients [36].

Thus, the battle with COVID-19 continues to date, and the secondary infections/ co-infections arising in COVID-19 result in overuse of the antimicrobials.

1.6 The need to reduce AMR to prevent an upcoming pandemic

A robust antimicrobial stewardship programme is recommended to reduce the rise in AMR and prepare or prevent the community from an upcoming pandemic due to AMR. In October 2015, the WHO launched the Global Antimicrobial Resistance and Use Surveillance System (GLASS), the first global collaborative effort to standardise AMR surveillance and is a crucial tool for infection control and prevention entities to use in developing policies. GLASS covers a group of pathogens (*Acinetobacter*, *E. coli*, *K. pneumoniae*, *N. gonorrhoeae*, *Salmonella*, *Shigella*, *S. aureus*, *S. pneumoniae*) which are essential for AMR surveillance [37]. Simultaneously, WHO launched its Global Plan on Antimicrobial Resistance (GAP) in 2015, where intending increased awareness of AMR; increased surveillance of AMR and antibiotic use; implement suitable infection control measures; implement antibiotic stewardship programmes; and increased economic investment to assist AMR reduction were among the declared goals [37].

Considering GAP, India launched the National Action Plan on AMR (NAP AMR) in 2017 [38]. Although promising, the NAP has yet to be fully implemented in India. Although a study in India indicated that antimicrobial stewardship program had a considerable favourable influence on reducing empirical antibiotic use and boosting targeted therapy, as evidenced by a significant increase in bacterial and blood culture rates [39]. In 2013 and 2014, the Indian Council of Medical Research and the National Centre for Disease Control (CDC) established AMR surveillance networks. These two organisations began a comprehensive study of current AMR practises in India in 2015, with cooperation from the CDC, intending to create new guidelines for reducing hospital-acquired infections [40]. Several national health authorities have recently recognised the importance of AMR in the environment, calling for standardisation of AMR guidelines, restrictions on the use of antibiotics as growth promoters in animal livestock, and pharmacovigilance, including prescription audits that include antibiotic usage in the hospital and community [41].

Since 2017, WHO data has shown an improvement in global surveillance infrastructure, and in 2019, a redesigned module to record antimicrobial usage to GLASS was launched; however, only nine countries have enrolled to date, none of which are from low/middle-income countries [37].

While health systems have traditionally focused on disease treatment, increasing cases of AMR and impending pandemics indicate that all health systems should adopt comprehensive preventative methods. By initiating prevention-oriented activities, the global health community can play a unique role in reducing mortality and morbidity due to AMR and pandemics. As a result, AMR stewardship initiatives should promote logical and prudent medicine use in persons, agriculture, and livestock to extend the efficacy of existing drugs [3]. Similarly, encouraging treatment adherence, averting suboptimum drug consumption practices like self-prescription, taking spare medicines from earlier prescribed doses, sharing leftover drugs with others, are highly recommended.

While multiple approaches are required to comprehend the complexity of AMR and implement an effective AMR stewardship programme, experts believe that behavioural science can aid in understanding and modifying essential public, patient, professional, and organisational behaviour. Qualitative approaches can be combined with a behavioural science approach to investigate impacts on behaviour and create and evaluate behavioural interventions [42].

As a result, there are a variety of successful measures for reducing antibiotic consumption and preventing a pandemic. Overall, AMR stewardship measures are vital instruments in the fight against multidrug resistance morbidity. Despite an increase in hospital stewardship programmes, it has to be seen whether their gains are sustainable in light of budget constraints and political support from above.

1.7 Further perspectives

The emergence of multidrug-resistant bacteria around the world is jeopardising our ability to treat infections and leaving us with less time to find new treatments. The fight for new antibiotics has been more difficult in recent years, as research resources and funds have been significantly shifted into producing COVID-19 treatments and vaccines [43]. In recent years, the AMR problem has garnered attention in several low/ middle-income countries. The most major achievement, however, has been establishing AMR surveillance for common human bacterial infections. Significant financial resources have hampered the implementation of comprehensive AMR national action plans. Limiting antibiotic usage in humans and agriculture will not be enough to manage and reverse AMR in low/ middle-income nations. To address the AMR problem, several major challenges in these sectors would necessitate a significant infrastructure revamp. An effective response will necessitate low-cost, long-term solutions to improve household and agricultural wastewater management, correctly handle pharmaceutical waste, and improve cleanliness. Long-term AMR solutions require the commitment of politicians and policymakers. Low/ middle-income countries should concentrate on enacting regulatory measures to prevent unauthorised antimicrobial use, AMR stewardship, and treatment guidelines for common infections, as well as long-term public awareness campaigns aimed at changing health-seeking behaviour. The current pandemic highlights how dangerous it is for an illness to spread quickly when there are no treatments available, especially when the most vulnerable people of society are disproportionately affected [4]. Despite the fact that the AMR epidemic has progressed slowly thus far, the loss of efficacy of essential antibiotics is increasing. Before the tipping point is reached, the emergence and spread of multidrug-resistant bacterial diseases must be stopped.

Bibliography

[1] Saira George, Fiorinda F Muhaj, Celine D Nguyen, and Stephen K Tyring. Part i antimicrobial resistance: Bacterial pathogens of dermatologic significance and implications of rising resistance. *Journal of the American Academy of Dermatology*, 2022.

[2] CDC. CDC launches networks to boost global response to antibiotic resistance — CIDRAP, 2021.

[3] Raskin Chaudhary, Shrawan Kumar Thapa, Jid Chani Rana, and Pradeep Kumar Shah. Surgical site infections and antimicrobial resistance pattern. *Journal of Nepal Health Research Council*, 15(2):120–123, 2017.

[4] Giorgia Sulis, Sena Sayood, and Sumanth Gandra. Antimicrobial resistance in low-and middle-income countries: current status and future directions. *Expert review of anti-infective therapy*, 20(2):147–160, 2022.

[5] Jim O'Neill. Tackling drug-resistant infections globally: final report and recommendations. 2016.

[6] World Health Organization et al. Blueprint for r&d preparedness and response to public health emergencies due to highly infectious pathogens. *Geneva, Switzerland: WHO*, 7, 2015.

[7] Govindaraj Vaithinathan Asokan and A Vanitha. Who global priority pathogens list on antibiotic resistance: an urgent need for action to integrate one health data. *Perspectives in public health*, 138(2):87–88, 2018.

[8] Bilal Aslam, Wei Wang, Muhammad Imran Arshad, Mohsin Khurshid, Saima Muzammil, Muhammad Hidayat Rasool,

Muhammad Atif Nisar, Ruman Farooq Alvi, Muhammad Aamir Aslam, Muhammad Usman Qamar, et al. Antibiotic resistance: a rundown of a global crisis. *Infection and drug resistance*, 11:1645, 2018.

[9] Longzhu Cui, Xiaoxue Ma, Katsuhiro Sato, Keiko Okuma, Fred C Tenover, Elsa M Mamizuka, Curtis G Gemmell, Mi-Na Kim, Marie-Cecile Ploy, N El Solh, et al. Cell wall thickening is a common feature of vancomycin resistance in staphylococcus aureus. *Journal of clinical microbiology*, 41(1):5–14, 2003.

[10] Rustam I Aminov. A brief history of the antibiotic era: lessons learned and challenges for the future. *Frontiers in microbiology*, 1:134, 2010.

[11] R Trastoy, T Manso, L Fernández-García, L Blasco, A Ambroa, ML Perez Del Molino, G Bou, R García-Contreras, TK Wood, and M Tomás. Mechanisms of bacterial tolerance and persistence in the gastrointestinal and respiratory environments. *Clinical microbiology reviews*, 31(4):e00023–18, 2018.

[12] Bram Van den Bergh, Maarten Fauvart, and Jan Michiels. Formation, physiology, ecology, evolution and clinical importance of bacterial persisters. *FEMS Microbiology Reviews*, 41(3):219–251, 2017.

[13] Rory Cave, Jennifer Cole, and Hermine V Mkrtchyan. Surveillance and prevalence of antimicrobial resistant bacteria from public settings within urban built environments: Challenges and opportunities for hygiene and infection control. *Environment international*, 157:106836, 2021.

[14] Gerardo Alvarez-Uria, Sumanth Gandra, and Ramanan Laxminarayan. Poverty and prevalence of antimicrobial resistance in invasive isolates. *International Journal of Infectious Diseases*, 52:59–61, 2016.

[15] Sunil Pokharel, Shristi Raut, and Bipin Adhikari. Tackling antimicrobial resistance in low-income and middle-income countries, 2019.

[16] James A Ayukekbong, Michel Ntemgwa, and Andrew N Atabe. The threat of antimicrobial resistance in developing countries: causes and control strategies. *Antimicrobial Resistance & Infection Control*, 6(1):1–8, 2017.

[17] Theodore Kelesidis, Iosif Kelesidis, Petros I Rafailidis, and Matthew E Falagas. Counterfeit or substandard antimicrobial drugs: a review of the scientific evidence. *Journal of Antimicrobial Chemotherapy*, 60(2):214–236, 2007.

[18] Peter Collignon and John J Beggs. Socioeconomic enablers for contagion: factors impelling the antimicrobial resistance epidemic. *Antibiotics*, 8(3):86, 2019.

[19] Feng Guo, Bing Li, Ying Yang, Yu Deng, Jian-Wen Qiu, Xiangdong Li, Kenneth MY Leung, and Tong Zhang. Impacts of human activities on distribution of sulfate-reducing prokaryotes and antibiotic resistance genes in marine coastal sediments of hong kong. *FEMS Microbiology Ecology*, 92(9):fiw128, 2016.

[20] Laura JV Piddock. The crisis of no new antibiotics—what is the way forward? *The Lancet infectious diseases*, 12(3):249–253, 2012.

[21] C Lee Ventola. The antibiotic resistance crisis: part 1: causes and threats. *Pharmacy and therapeutics*, 40(4):277, 2015.

[22] Tingting Jiang and Xiang-Sheng Chen. Outcome impacts due to pathogen-specific antimicrobial resistance: A narrative review of published literature. *International Journal of Environmental Research and Public Health*, 17(4):1395, 2020.

[23] Christopher A Ohl and Vera P Luther. Health care provider education as a tool to enhance antibiotic stewardship practices. *Infectious Disease Clinics*, 28(2):177–193, 2014.

[24] Michalis Karvelas, Athanasios Katsoyiannis, and Constantini Samara. Occurrence and fate of heavy metals in the wastewater treatment process. *Chemosphere*, 53(10):1201–1210, 2003.

[25] David W Nelson, John E Moore, and Juluri R Rao. Antimicrobial resistance (amr): significance to food quality and safety. *Food quality and safety*, 3(1):15–22, 2019.

[26] Michelle MC Buckner, M Laura Ciusa, Richard W Meek, Alice R Moorey, Gregory E McCallum, Emma L Prentice, Jeremy P Reid, Luke J Alderwick, Alessandro Di Maio, and Laura JV Piddock. Hiv drugs inhibit transfer of plasmids carrying extended-spectrum β-lactamase and carbapenemase genes. *Mbio*, 11(1):e03355–19, 2020.

[27] Emily Mabonga, Rosalind Parkes-Ratanshi, Stefan Riedel, Sheila Nabweyambo, Olive Mbabazi, Chris Taylor, Charlotte Gaydos, and Yukari C Manabe. Complete ciprofloxacin resistance in gonococcal isolates in an urban ugandan clinic: findings from a cross-sectional study. *International journal of STD & AIDS*, 30(3):256–263, 2019.

[28] SM S Wilmore, K Kranzer, A Williams, B Makamure, AF Nhidza, J Mayini, T Bandason, J Metcalfe, MP Nicol, I Balakrishnan, et al. Carriage of extended-spectrum beta-lactamase-producing enterobacteriaceae in hiv-infected children in zimbabwe. *Journal of medical microbiology*, 66(5):609, 2017.

[29] Keiichi Hiramatsu. Vancomycin-resistant staphylococcus aureus: a new model of antibiotic resistance. *The Lancet infectious diseases*, 1(3):147–155, 2001.

[30] Bradley J Langford, Miranda So, Sumit Raybardhan, Valerie Leung, Duncan Westwood, Derek R MacFadden, Jean-Paul R Soucy, and Nick Daneman. Bacterial co-infection and secondary infection in patients with covid-19: a living rapid review and meta-analysis. *Clinical microbiology and infection*, 26(12):1622–1629, 2020.

[31] Ting-Kuang Yeh, Zong-Hao Li, Yao-Ting Huang, and Po-Yu Liu. Covid-19 associated bacteremia with chryseobacterium indologenes co-harboring blaind-2, blacia and blaccra. *Infection and Drug Resistance*, 15:167, 2022.

[32] Nesrine A Rizk, Rima Moghnieh, Nisrine Haddad, Marie-Claire Rebeiz, Rony M Zeenny, Joya-Rita Hindy, Gabriella Orlando, and Souha S Kanj. Challenges to antimicrobial stewardship in the countries of the arab league: Concerns of worsening resistance during the covid-19 pandemic and proposed solutions. *Antibiotics*, 10(11):1320, 2021.

[33] Anas Baiou, Abdurrahman A Elbuzidi, Dana Bakdach, Ahmed Zaqout, Khalid M Alarbi, Awadh A Bintaher, Maisa MB Ali, Anam M Elarabi, Gawahir AM Ali, Joanne Daghfal, et al. Clinical characteristics and risk factors for the isolation of multi-drug-resistant gram-negative bacteria from critically ill patients with covid-19. *Journal of Hospital Infection*, 110:165–171, 2021.

[34] Takwa E Meawed, Sherweet M Ahmed, Sherif MS Mowafy, Ghada M Samir, and Reham H Anis. Bacterial and fungal ventilator associated pneumonia in critically ill covid-19 patients during the second wave. *Journal of infection and public health*, 14(10):1375–1380, 2021.

[35] Vikas Saini, Charu Jain, Narendra Pal Singh, Ahmad Alsulimani, Chhavi Gupta, Sajad Ahmad Dar, Shafiul Haque, and Shukla Das. Paradigm shift in antimicrobial resistance pattern of bacterial isolates during the covid-19 pandemic. *Antibiotics*, 10(8):954, 2021.

[36] Katarzyna A Kazimierczak, Karen P Scott, Denise Kelly, and Rustam I Aminov. Tetracycline resistome of the organic pig gut. *Applied and environmental microbiology*, 75(6):1717–1722, 2009.

[37] World Health Organization et al. Health care-associated infections fact sheet. *World Health Organization*, 4, 2015.

[38] Jaya Ranjalkar and Sujith J Chandy. India's national action plan for antimicrobial resistance–an overview of the context, status, and way ahead. *Journal of family medicine and primary care*, 8(6):1828, 2019.

[39] Rahul Garg, Gyanendra Singh, Shweta Kumar, Mamta Verma, Lily Podder, Vaibhav Ingle, Abhishek Singhai, T Karuna, Saurabh Saigal, Kamini Walia, et al. Impact of an anti-microbial stewardship program on targeted antimicrobial therapy in a tertiary care health care institute in central india. *Cureus*, 13(10), 2021.

[40] Soumya Swaminathan, Jagdish Prasad, Akshay C Dhariwal, Randeep Guleria, Mahesh C Misra, Rajesh Malhotra, Purva Mathur, Kamini Walia, Sunil Gupta, Aditya Sharma, et al. Strengthening infection prevention and control and systematic surveillance of healthcare associated infections in india. *bmj*, 358, 2017.

[41] Neelam Taneja and Megha Sharma. Antimicrobial resistance in the environment: The indian scenario. *The Indian journal of medical research*, 149(2):119, 2019.

[42] Aleksandra J Borek, Marta Santillo, Marta Wanat, Christopher C Butler, and Sarah Tonkin-Crine. How can behavioural science contribute to qualitative research on antimicrobial stewardship in primary care? *JAC-Antimicrobial Resistance*, 4(1):dlac007, 2022.

[43] David Adam. Scientists fear that'covidization'is distorting research. *Nature*, 588(7838):381–383, 2020.

2

Diversity in the Development and Transmission of AMR

Riya Joshi

Symbiosis School of Biological Sciences, Symbiosis International (Deemed University) Pune, 412115, India

Agnita Roychowdhury

Symbiosis School of Biological Sciences, Symbiosis International (Deemed University) Pune, 412115, India

Sunil D. Saroj

Symbiosis School of Biological Sciences, Symbiosis International (Deemed University) Pune, 412115, India

CONTENTS

2.1 Introduction

Antibiotics are the wonder drugs that have served humans in the battle against a vast variety of microbial infections. For many years, these antibiotics have been used extensively for therapeutic indications. The misuse and overuse of these antimicrobial agents is a major driving force heading towards resistance. Antimicrobial resistance (AMR) is a serious global concern of the 21st century that has emerged in the environment, humans, and animals [1]. Excessive and imprudent consumption of antibiotics in different fields such as medical, agriculture, and veterinary are the possible root cause of AMR development and its worldwide spread. Antimicrobials are the most commonly used drugs in human medicine; however, up to 50 percent of the prescribed antimicrobials are considered unnecessary [2]. Inappropriate selling of antibiotics, poor sanitation practices, discharge of antibiotics or their residues through manure or faces, and release of industrial waste into the environment have resulted in worsening of AMR problem. There is an interrelated relationship between these contributing factors and the spread of AMR through the environment. The antibiotics discovery has been a revolt in the field of therapeutics and medicine. However, large-scale production and consumption have caused widespread effects on the mankind and microbial biosphere. Studies have shown a rapid 36 percent increase in global antibiotic consumption by humans from 2000 to 2010 [3]. Antibiotic intake by animals follows a similar pattern. Such a continually rising amount of antibiotic consumption overwhelms the process of natural production, constantly elevating the selection pressure on a diverse population of bacteria in all the exposed environment [4].

Bacterial resistance evolves as a result of natural selection. Resistance develops genetically, either through mutation or through the acquisition of new resistance genes via genetic exchange processes[5]. Antibiotics have long been known to enhance pathogen resistance mechanisms via horizontal gene transfer. Resistance to third-generation antibiotics such as polymyxins, cephalosporins, and carbapenems is on the rise [6]. The resistance to third generation cephalosporin is through uptake of extended-spectrum β-lactamase (ESβL) genes. According to the estimates, about 14 percent of the human global population is colonized by ESβL which is growing at a 5.4 percent rate annually [7]. Due to increasing microbial resistance to vancomycin, WHO has urged to finance research for novel antimicrobial molecule development against these priority pathogens. Concerns regarding carbapenem and polymyxin are also rising that take control over broad-spectrum Gram-negative bacteria. One such incidence is of the New Delhi Metallo-β-lactamase type 1 (NDM-1) gene which was reported in 2008 in a clinical patient [8]. The prevalence of non-antibiotic compounds in environmental samples is also accountable for resistant pathogen selection. Nickel, zinc, copper, and mercury-mediated antimicrobial-resistant gene selection has also been well reported and studied [9, 10]. The world's leading health surveillance systems and agencies trace AMR as of outmost global issue due to the rapidly expanding human and animal population, increased globalization, international trade, rising demand for food, and easy access of antimicrobial compounds to developed and developing counties. A recent report highlights the lack of necessary knowledge on existing antimicrobials and their future applications. Hence, this chapter briefly highlights the emergence of antimicrobial resistance, transmission, the mechanism involved in antimicrobial resistance, and modes of AMR dissemination.

2.2 Antimicrobial resistance

The discovery of antibiotics is observed to be a key milestone of the 20th century. As a consequence of antibiotic therapy, the mortality and morbidity rates of infectious diseases have been lowered significantly[11]. The antibiotic era is considered to have begun in the 1900s, starting with the dark ages, followed by the golden age when the antibiotics in use today were discovered, followed by the lean ages when the use of antibiotics was improved by appropriate dosing [5]. Antimicrobial substances are known to kill or inhibit the growth of microorganisms that are either produced naturally or synthetically. Antimicrobials are the broad umbrella

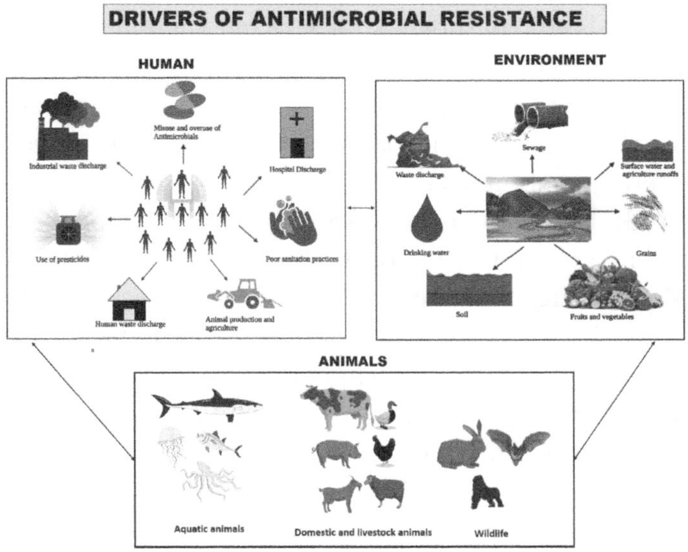

FIGURE 2.1
Schematic overview of various driver of AMR: The challenge of antimicrobial resistance emergence and transmission is driven by several factors such as human, animal and environmental.

that consists of antibiotics, antivirals, antifungals, and antiparasitics. Antimicrobial agents are being used in symptomatic as well as prophylactic treatment. In addition to its administration to maintain human health, it is also being used in animal and plant health to prevent diseases and promote growth. In fact, the therapeutic use of antibiotics in human health accounts for less than 50 percent of the total antibiotics produced[5]. As a result, microorganisms are repeatedly exposed to antimicrobial agents, leading to the development of resistance against them as a result of Darwinian selection. This is a natural evolutionary process[12]. Many bacterial pathogens that are known to cause epidemics of human health have evolved to extensively drug-resistant, multidrug-resistant and pandrug-resistant strains, for example, the ESKAPE group of pathogens. This has resulted in superbugs, with increased morbidity and mortality, as well as super-resistant strains, having enhanced virulence genes and transmissibility [5]. These pathogens have been listed by the WHO as priority pathogens for which new antimicrobial drugs need to be developed. Infections due to multidrug-resistant pathogens have taken the annual death toll to approximately 700,000 and if a potent action plan is not put into operation, by the year 2050 annual death rate can reach up to 10 million deaths per year that is more than the deaths due to cancer and diabetes combined [4]. In addition to this, according to the World Bank, by the year 2050 AMR will increase poverty rates by causing a 1 percent worldwide GDP loss and a 5-7 percent GDP loss in developing countries [12]. With the rise in AMR, the absence of novel drugs, and the failure of genome-based methods to battle with this problem, we appear to be approaching the disenchantment stage, where handwashing will be the only viable preventative measure [5]. While the development of new antimicrobial drugs is essential, it is also pertinent to have a comprehensive understanding of the drivers, origin, mechanisms, and dissemination of AMR in order to combat this global issue.

2.2.1 Origin of resistance

The resistance shown by bacteria is never uniform and the degree of resistance varies across bacterial species [13]. They continuously adapt and evolve in response to stress. Similarly, in response to antimicrobial stress, bacteria will defend itself by gaining resistance. This resistance exhibited by the pathogen can have the following origins:

2.2.1.1 Natural resistance

Natural resistance can be intrinsic or induced. Intrinsic resistance is conferred by the resistance genes that are always expressed, whereas induced resistance is observed only in the presence of antibiotics by the expression of AMR genes that are indigenous to the bacteria but are not constitutively expressed [13].

2.2.1.2 Acquired resistance

The resistance is exhibited either by the virtue of the acquisition of genetic material that confers resistance, like horizontal gene transfer (HGT), or by mutations to the chromosomal DNA. This acquired resistance might be permanent or temporary. The most prevalent mode of acquiring resistance genes is by plasmid-mediated transmission, whereas bacteriophage-mediated transmission is the least common. Mutations on the other hand can be caused by integrins or insertional sequences or in response to some external stress like exposure to UV or starvation. More often than not, these mutations occur at a cost. For example, *S. aureus* gains resistance against methicillin at the price of a significant reduction in its growth rate [10].

2.2.1.3 Adaptive resistance

Adaptive resistance is defined as a temporary increase in the ability of a bacterium to survive an antibiotic insult due to alterations in gene and/or protein expression as a result of exposure to an environmental trigger [14]. This includes resistance due to exposure to sub-lethal concentrations of antibiotics, the presence of environmental cues, or the formation of biofilms [14].

2.3 Drivers of AMR - Emergence and Spread

Antimicrobial resistance (AMR) has emerged as a prime example of how bacterial populations can rapidly adapt and react to selective pressure due to the widespread use of antimicrobial drugs in human and veterinary medicine and animal husbandry [15]. We have seen not only the birth of a slew of new resistance mechanisms over the last few decades but also their diffusion across entire bacterial populations and ecological niches [16]. Researchers and stakeholders are increasingly concerned that the environment serves as an AMR reservoir and plays a crucial role in the spread of antimicrobial resistance genes (ARGs) [17]. Antimicrobial medication use in healthcare, agriculture/livestock, and the environment, as well as antibiotic residues produced from various residential settings, are all major contributors to the development of antibiotic-resistant bacteria and associated ARGs [17]. Diverse pathways or hotspots, such as hospital effluent, agricultural waste, or wastewater treatment plants, allow resistant microbes and their resistance genes to enter the soil, air, water, and sediments [15]. The discovery of high-risk situations that are responsible for resistance evolution and propagation is at the heart of global mitigation initiatives. As a result, sanitation, access to clean water, availability of high-quality antimicrobials and diagnostics, infection control, travel, and migration all have an impact on AMR transmission (Figure 2.1).

Considerable research data has indicated the drivers conferring antimicrobial resistance, which include: 1. Antimicrobials, such as antivirals, antibacterial, antifungal, and antiparasitic drugs; 2. Biocides: Surfactants and disinfectants; and 3. Heavy metals. Additionally, resistance to xenobiotics like octanol, hexaneoctanol, hexane and toluene; as well as natural chemicals that are plant-derived has been found to pick for resistance genes [14]. For hazardous compounds such as solvents, biocides [18], heavy metals [19], and antibiotics, the co-selection of resistance genes have been described. Co-selection can take two forms: First being co-resistance, in which one gene assists the choosing of another gene that does not usually offer a selective advantage to the interesting compound [20]; and second is cross-resistance, in which one resistance gene defends against a variety of toxic chemicals [21]. The principal environmental transmission mechanisms for these drivers include (i) aquaculture systems, (ii) sewage sludge and animal manure spreading, and (iii) wastewater from industrial and municipal cooperation. Some of these drivers are listed below.

2.3.1 The driver of resistance: Antibiotic and antibiotic resistance genes

The driver of resistance: Antibiotic and antibiotic resistance genes are as follows.

2.3.1.1 Antibiotic consumption in human

Antibiotics have been used to treat a variety of human infectious diseases ever since the discovery of penicillin. Antibiotics are the most frequently prescribed drugs in human medicine; however, up to 50 percent of prescribed antimicrobials are regarded unnecessary. Antibiotics are still widely used without a prescription in many countries outside of Europe and North America, accounting for 19 percent of total use and even 100 percent in some situations. Multidrug-resistant (opportunistic) pathogens like *Acinetobacter baumannii* and *Pseudomonas aeruginosa* can thrive in places where other bacteria cannot by disrupting the microbiota of the host [22]. The abuse and misuse of antibiotics lead to the selection of such resistant strains and is considered as a driving force to resistance [23]. The antibiotic prescription in humans is found to be excessive in the inpatient settings and overall prescription is highest in the primary care community. Several confounding factors contribute towards the link between antibiotic use by humans and resistance, which includes: pathogen-host interaction, the emergence of resistant clones, pathogen drug interaction, mutants in the pathogen, cross-resistance, and transmission of pathogen among environment, animals, and humans. Moreover, at the human population level, factors such as vaccine administration rate, migrations, sanitation, healthcare systems, and population density are a major influence in the resistance prevalence [1]. Role of humans in the transmission of antibiotic resistance: The transmission dynamics have helped improve the understanding of the way in which human contributes to the escalating spread of AMR. The common route of transmission in the community is the fecal-oral transmission that occurs often through negligence in sanitation and plays a beneficial role for resistant *Enterobacteriaceae* particularly. Another mode of transmission occurs by sexual encounters; *N. gonorrhoea* is the core group of pathogens that have contributed to the dissemination of these antimicrobial-resistant clones [24]. The inability to speedily identify microorganisms resistant to antimicrobials with appropriate tests can further contribute towards the transmission of resistance in healthcare settings. The human microbiota has rapidly acquired AMR to *Enterobacteriaceae* in the span of ten years. In many parts of the world, the ESβL-positive *Enterobacteriaceae* carrier rate in the gut is approximately 50 percent, and its travel is clearly linked with high chances of gut colonization [21]. Another worrying factor is the worldwide spread of resistance mechanism carbapenem and also between organisms with OXA-48, NDM, and KPC enzymes are of great concern [8, 25, 26]. In order to reduce human-human transmission, several population-based strategies can be adopted that can alter the dynamics of transmission of pathogens such as mass drug administration and targeted vaccinations. The vaccines can result in the elimination or reduction of infection, in turn reducing the requirement of antimicrobial treatment [27].

2.3.1.2 Antibiotic consumption in animals

According to The State of the World's Antibiotics, animal husbandry uses two-thirds of all antibiotics produced globally each year (65,000 tonnes out of 100,000 tonnes) [3]. Even within high-income countries, there can be large differences in veterinarian antimicrobial use in food-producing animals. The sole purpose of veterinary drugs was for the prevention of animal infections, however, they are being employed for various other purposes like growth promoters, preventative agents, and feed additives. After China, the United States, and Brazil, India is the fourth largest manufacturer of antimicrobials for animals. Antibiotic residues have been found in food-animal products such as fowl, beef, and milk [28]. According to the Food and Drug Administration's annual summary report on the sale and distribution of antimicrobials in food-producing animals, consumption of medically significant antimicrobial medicines has increased by 12 percent.

Animals, like humans, release 30–90 percent of the consumed antibiotics into manure and urine. Antibiotic residues in agricultural waste are elevated due to the high use of antibiotics on livestock farms, contaminating the environment with antibiotic-resistant bacteria [29]. These scenarios facilitate antibiotic resistance gene selection and transfer within the environment. These resistance genes are further transferred to other environments and humans through their entry into the food supply and waste products [30, 29]. *Salmonella* spp. and *E. coli* O157:H7 have been found in manure, compost, soil, and water samples. Salmonella was found to be detected in 6 of 380 pig farm manure samples and 6 of 96 water samples from leafy green farm irrigation and non-irrigation sources [31]. The exploitation of antimicrobials in the veterinary and medical professions has resulted in increased AMR as a result of selection pressure against antimicrobial agents.

Role of animals in the transmission of resistance: The possibility for microbe AMR transmission from animals to humans, with the interrelation between antimicrobial growth promoter usage in farm animals and resistant bacteria transmission, was identified in the 1960s [32]. Mobile genetic elements and bacteria that confer resistance stay in the feces and on the skin of animals which can be further transmitted to different bacteria and to humans [33]. This interweaving of the human and animal ecosystem is extended to opportunistic and commensal pathogen, like species such as *Staphylococcus aureus*, *E.coli*, and Enterococci.

2.3.1.3 Antibiotic pathway in environment

AMR can be transmitted to humans, animals, and the environment through a variety of methods. The ecosystem acts as a link between diverse animal compartments, soil, water, sand, and sewage. The environment serves as a reservoir for mobile genetic materials to mix and propagate to other areas of the body, as well as human and animal hosts [34]. Antibiotics are discharged in a variety of ways, including municipal and hospital waste, animal husbandry, the manufacturing industry, runoff from agricultural fields containing livestock manure, and landfill leachates [20]. Antibiotics have a half-life ranging from hours to hundreds of days, however, antibiotic residues are considered environmental pollutants [34]. Antibiotic use is increasing, putting a lot of pressure on natural microbial ecosystems and humans [12][12]. Unused or expired antibiotics, like other pharmaceutical items, are thrown away and end up in the groundwater or aquatic system during the wastewater treatment process, where they cannot decompose. By horizontal gene transfer or modifying target sites, a relatively low concentration of these discarded antibiotics can promote selective pressure[12]. Such resistance genes are primarily found in the human gastrointestinal tract [35]. Similarly, multiple active antibiotic resistance genes were found in 70 percent of Gram-positive and 65 percent of Gram-negative bacterial isolates from a natural cave microbiome, suggesting that multiple resistance is an environmental standard rather than an exceptional response to elevated anthropogenic antibiotic stress. Antibiotic-resistant bacteria and ARGs were found at higher concentrations in a variety of wastewater samples, including municipal wastewater, sewage, and wastewater treatment plant influents and effluents, and in agricultural and industrial effluent [8, 36]. The antibiotic through various sources going through various treatment stages and is then released into the environment. Antimicrobial waste is generated in hospitals and other healthcare institutions in two ways: indirectly through patient secretions and directly through wasted drugs. Novel ARGs, concentrated in activated sludge and effluent wastewater, have been found to confer resistance to a variety of antibiotics, including aminoglycosides, fluoroquinolones, and β-lactams [37]. ESβL-producing *E. coli*, MRSA, and resistant enterococci have been discovered in activated sludge, with the *vanA* and *mecA* genes giving resistance to vancomycin and β-lactams, respectively [21]. India and China are the two countries contributing the most to the worldwide rise of AMR, necessitating immediate observation. Antibiotic-producing environmental microbes, unlike clinical microbes, show a higher level of inherent resistance that appears to be independent of selective pressure and is not acquired through horizontal gene transfer. The possibility of transferring inherent resistance found in ambient microbes to human-adapted diseases is growing, and it has become a major worry in hospital settings.

2.3.2 Drivers of resistance: Biocides

Biocides are the disinfectants that are commonly used in hospitals, cosmetics, household cleaning products, hospitals, wipes, furniture preservatives, farmyards, and a variety of industrial processes [38]. Chlorhexidine, formaldehyde, triclosan, quaterium ammonium compounds, and ethanol are examples of common biocides [39]. Six distinct biocides have been demonstrated to have MICs ranging from 0.4 to 1000 mg/L against sixteen different bacteria (including *Pseudomonas aeruginosa*, *E. coli*, *Staphylococcus aureus*, and *Citrobacter* spp.). Antibiotic resistance has been demonstrated to be selected by triclosan, as well as other biocides such as chlorhexidine and QACs [40]. Sub-lethal concentrations of biocides have been demonstrated to select for common mutations that confer clinically meaningful antibiotic resistance in the same way that sub-lethal doses of antibiotics select for ARGs [18]. As previously mentioned, several resistance pathways are shared between biocides and antibiotics, allowing for ARG co-selection [29]. *S. aureus* co-resistance to the biocide benzalkonium chloride and the antibiotic oxacillin, allowing the resistant strain to tolerate the antibiotic eight times better than the susceptible wild-type strain. Because these genes are frequently co-located with ARGs, they have greater resilience and mobility, allowing them to spread quickly through microbial communities and pathogens [41].

2.3.3 The driver of resistance: Heavy metal

The major sources of heavy metals to WWTP are through household effluent, business effluent, drainage water, traffic-related emissions, and atmospheric deposition [42]. Metal nanoparticles such as silver, titanium, and copper are widely used in disinfectants, food, textiles, hospital products, household, and industrial products [43]. Metals like As, Pb, Cu, Zn, and Cd have been used as pesticides, fertilizer, nutrient supplements, and fungicides in agriculture and aquaculture as well as animal growth promoters [41]. Metal and antibiotic resistance share certain functional and structural resistance systems, like (1) Drug and metal efflux of metals such as Ni, Cu, Zn, Co, Cd, and As is similar to antibiotics such as β-lactams, tetracycline, and chloramphenicol; (2) Metal alteration of As, Hg is similar to the drug alteration of antibiotics: chloramphenicol and β-lactams; (3) Reduced membrane permeability of metals such as As, Ag, Zn, Cu, Mn, Co is similar to antibiotics like beta-lactams, tetracycline, ciprofloxacin, and chloramphenicol; (4) Cellular target alteration of metals such as Cu, Hg, Cu similar to antibiotics such as rifampicin, ciprofloxacin, trimethoprim, and β-lactams; (5) Metal of Zn, Cd, Cu similar to drug sequestration of coumermycin [44]. Substantial literature is available on species toxicity and mobility in sediment, aquatic, and soil ecosystems [45, 46]. Metal resistance genes have been found on plasmids, and bacteria with metal resistance genes have been found to carry ARGs more frequently than bacteria without metal resistance genes. Because the combined exposure of antibiotics, biocides, and metals exceeds the minimum selection concentration for some shared resistance mechanisms, the desired goal of limiting resistance selection within soil habitats necessitates revisiting the environmental thresholds of sludge and manure land spreading [47].

2.4 Molecular mechanisms of AMR

Antibiotics have a specific mechanism of action and overtime microorganism and have also developed specific molecular mechanisms against each class of drug. A distinct mechanism for resistance is present for all antibiotics in use and this has made treatment of infections challenging. Over the years, the study of these molecular mechanisms has increased owing to the advancements in the field of genomics and systems biology. A comprehensive understanding of these mechanisms is crucial in the discovery of novel therapeutics [41] (Figure 2.2). The molecular mechanism of AMR displayed by microorganisms is as follows.

2.4.1 Reduced drug uptake

Bacteria possess an innate resistance by limiting the uptake of certain antibiotics. In Gram-negative bacteria, the LPS (lipopolysaccharide) layer serves as a barrier against some hydrophilic antibiotics. Gram-positive bacteria lack this outer membrane and hence cannot effectively prevent the uptake of drugs. However, *S. aureus* is transiently resistant to vancomycin by the production of a thick cell wall and Enterococci have intrinsic resistance to aminoglycosides as they are polar molecules that cannot penetrate the cell wall [9] (Figure 2.2). Another mechanism to reduce the drug uptake is the reduction in the number of porin channels or changes in porin selectivity. *Enterobacteriaceae* is known to reduce the number of porin channels or in some cases completely stop its production to obtain resistance against carbapenems [21]. Meanwhile, *N. gonorrhoeae* have mutations within the porin channel that provides resistance against β- lactams and tetracyclin [21]. Bacteria form biofilms on abiotic and biotic surfaces. These biofilms are also known to provide an adaptive resistance to the bacteria. It causes slow or incomplete penetration of antimicrobial drugs and contributes to increased resistance. In *P. aeruginosa* biofilm it was observed that the biofilm acts as a barrier to physical diffusion and slows the penetration of the antibiotic giving *P. aeruginosa* enough time to adapt to the antibiotic [48].

2.4.2 Drug target modification

The targets of the antimicrobial agents can be protected and modified in order to provide resistance. For resistance against β- lactams in Gram-positive bacteria, the PBP (penicillin-binding protein) is modified to have either a decreased production with regular drug-binding capacity or an increased production with decreased drug-binding capacity [49] (Figure 2.2). Gram-negative bacteria are known to have intrinsic resistance against the glycopeptides and lipopeptides due to the presence of the LPS layer. However, in Gram-positive bacteria like Enterococci and *S. aureus* resistance against these drugs is acquired. The VRE and the MRSA strains have emerged due to the acquisition of resistance genes [49]. Resistance against drugs that target the ribosomal subunit is achieved by mutation or methylation of the ribosomal subunit reducing the ability of the drug to bind to the ribosome. Mutations that cause modifications in the structure of gyrase and topoisomerase interfere with the ability of fluoroquinolones to bind to DNA gyrase and topoisomerase. Sulphonamides and trimethoprim bind to the active site of enzymes like dihydropteroate synthase and dihydrofolate via competitive inhibition and affect the growth of the bacterial cell. Hence mutations in the enzymes near the active site cause structural changes to the enzyme preventing the binding of the drug [49]. This resistance against antibiotics does not always necessitate the mutation in the genes that encode for the drug target. Target protection is another very effective mechanism of resistance. For example, the 16S rRNA can be methylated by the erythromycin ribosome methylase genes, protecting the drug target and altering the binding site [50]. Another example is the methylation of A2503 in the 23S rRNA by chloramphenicol–florfenicol resistance (CFR) methyltransferase that provides resistance against a wide range of antibiotics[27]. Bacteria can evolve to produce new targets that perform comparable biochemical functions to the original target but are not inhibited by the antimicrobial molecule. Significant clinical examples include methicillin resistance in *S. aureus* and vancomycin resistance in Enterococci [26]. Finally, by the overproduction of the target, the bacteria can bypass the metabolic process that the drug inhibits [51].

2.4.3 Inactivation/ degradation of drugs

Inactivation of drugs takes place by the transfer of a chemical group to the drug. For example transfer of the acetyl group to aminoglycosides inactivates them, rendering them ineffective (Figure 2.2) [49]. Aminoglycosides are specifically susceptible to this as they are large molecules with exposed hydroxyl and amide groups. Acetyltransferases, phosphotransferases, and nucleotidyltransferases are the three main classes of aminoglycoside modifying enzymes. Meanwhile, certain enzymes produced by the organisms like β-lactamases function by hydrolyzing the β- lactam ring of the drug, thereby degrading them. The initial β-lactamases were functional against first-generation β-lactams, then came the extended-spectrum β-lactamases (ESβLs) that were functional against oxyimino-cephalosporins. This led to the increased use of carbapenems in treatment that was followed by the isolation of clinical isolates that produced carbapenemase [49].

2.4.4 Efflux pump activation

Solutes, antimicrobial drugs, signaling molecules, and metabolites are pumped out of the cell by efflux pumps. Through efflux pumps, bacterial cells maintain their internal environment. In Gram-negative bacteria, drug efflux is a significant mechanism of resistance (Figure 2.2) [22]. The genes for efflux pump expression are present in the chromosomal DNA of the bacteria. Some of these genes are constitutively present while some are expressed only under the presence of an external stimulus [49].

The overexpression of efflux pumps leads to increased levels of resistance against antibiotics routinely used in clinical settings. Although all bacteria have numerous genes for MDR efflux pumps on their chromosomes, some of these genes have been mobilized onto plasmids and these genes are now being transferred to other bacteria, conferring resistance to them [49].

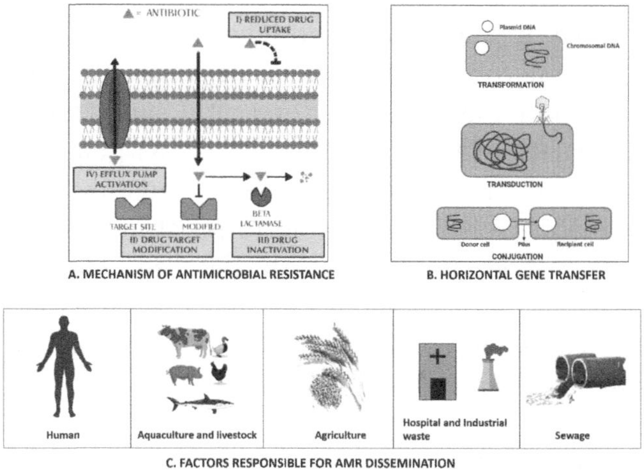

FIGURE 2.2
Antimicrobial resistance dissemination and resistance mechanisms (A) Schematic overview of the molecular mechanisms of the antimicrobial resistance as seen in gram negative bacteria, (B) Genetics of antimicrobial resistance dissemination by horizontal gene transfer, (C) Factors governing the dissemination of antimicrobial resistance genes.

2.5 Diversity in mode of dissemination

The environmental dispersion of antibiotics and the dissemination of antimicrobial-resistant genes have been linked to an increase in the number of resistant pathogens and commensal bacteria. Furthermore, antibiotics may cause changes in microbial community diversity in the soil as a result of their spread in the environment. Antibiotics are utilized in animal production all over the world, raising the danger of antimicrobial resistance (AMR) dissemination. Antibiotics, when used as preventative therapies, can boost selective pressure, promoting the development of antibiotic-resistant microorganisms [48]. To decrease the use of antibiotics in animal husbandry, better livestock and waste management methods, diets, proximity between animals, waste treatment, usage of additives, and operational conditions should be implemented [52]. Multiple mechanisms exist for antimicrobial resistance genes to enter and survive in the ecosystem. They disseminate across the soil, crops, gut microbial populations of wild and livestock animals, as well as humans [53, 54]. Horizontal gene transfer (HGT) using mobile genetic elements (MGEs) such as phages, plasmids, transposons, or integron gene cassettes spreads antimicrobial resistance genes[55]. Genes for antibiotic resistance are frequently found on mobile genetic elements (MGEs) [16]. The main source of MGE persistence and dissemination has been discovered as the horizontal transmission of MGEs to a wide spectrum of bacteria (pathogens and human commensals included).

Reservoir of AMR

To better understand how antibiotic resistance spreads, it is important to map the resistome of different habitats and determine to what extent these environments can operate as a reservoir for the spread of ARGs to bacterial pathogens. Several studies have employed different ways to sample the resistome of the various environments including soil, wastewater, and human and animal gut microbiota in recent years, and there has been an increase in interest in this issue [36, 9]. Since then, it has become obvious that ARGs, particularly clinically significant ones, are spread widely in such environments [52]. A growing number of such studies have discovered a large number of previously unknown ARGs in environments such as activated sludge [56], soil [53], the microbiota of animals [54], and that of humans [35]. Several examples of ARGs on MGEs in human pathogens that appear to have evolved from those reservoirs support the idea that commensals and the environment are major reservoirs for resistance. The blaCTXM genes, for example, have become the most common cause of extended-spectrum-lactamases (ESβLs) in *Enterobacteriaceae* globally and a major source of clinical treatment issues [57]. The chromosomal DNA of numerous environmental *Kluyvera* species was found as the likely source of these genes, from which they spread rapidly to various bacterial species.

The autochthonous bacteria comprises resistors or antibiotic resistance gene environmental reservoirs, which subsequently can get transferred to another pathogen via horizontal gene transfer.

For the horizontal dissemination process of antimicrobial resistance gene, three mechanisms are known to occur:

2.5.1 Transformation

Natural bacterial transformation is a well-known and explored mechanism involving the uptake, integration, and expression of naked fragments of DNA from the environment. For the process of transformation to occur, the following conditions should be fulfilled [58].

- The DNA should be present in the extracellular environment.
- The recipient bacteria should be competent.
- The translocated DNA should be stabilized.

Some bacteria are considered naturally competent like *Neisseria* spp. whereas, other bacterial species are capable of transformation by developing competence under certain specified conditions including the presence of autoinducers or peptides, stressful conditions, and nutritional status [59]. Furthermore, studies have proven that antibiotic exposure induces competence in many bacteria species. Hence, antibiotics will pick for the resistant strains and trigger the transformation. For instance, in Streptococci, the PBP encoding penicillin-binding protein with low affinity for beta-lactam antibiotics is the resultant of gene transfer from interconnected penicillin-resistant species which has further assisted in the dissemination of penicillin resistance among different *Streptococcus* spp [60, 50]. Mosaic variations of the genes *parC, parE,* and *gyrA* have been found to be easily transformed between *Streptococcus oralis, S. pneumonia*, and *Streptococcus mitis*, as well as between *Streptococcus pyogenes* and *Streptococcus dysgalactiae* [61]. Mobile genetic elements such as integrons, transposons, and gene cassettes have the capability of efficient dissemination between species, despite the genetic level relatedness. Streptococcal species are also shown to exchange transposons by transformation as well as conjugation [55]. This collectively indicates the broad capacity of transformation for the horizontal dissemination of resistance genes between various divergent species.

2.5.2 Transduction

The bacteriophages are important for creating a bacterial microbiome in the environment. Bacteriophages can transfer genes that are beneficial to their microbial hosts by specialized or broad transduction, enhancing their own survival and spread [62]. Transduction is the process where genes picked up during infection of another bacterium are sometimes carried by these bacteriophages and these genes are inserted into the new bacterial host's DNA. Bacteria can transfer DNA sequences ranging from MGEs like transposons, plasmids, and genomic islands to chromosomal DNA [63, 51]. Several studies have documented the role of bacteriophages in transfer or mobilization of ARGs for bacterial species, which includes: the beta-lactamase gene carriage by bacteriophage in *Salmonella* spp and *E. coli*; transduction of tetracycline, erythromycin, and other multiple resistance among *S. pyogenes* strains; the transfer of gentamycin and tetracycline resistance gene between Enterococci and the transfer of resistant plasmids in the MRSA [64, 65, 66]. ARGs in bacteriophages were found in the respiratory tract DNA of cystic fibrosis patients in another study [67]. The antibiotic treatment had resulted in increased ARGs number within the intestinal phagosome and increased interaction between bacteria and phages. Furthermore, several studies have detected ARGs presence in bacteriophages from animal waste, human fecal samples, wastewater samples, wastewater treatment plants, and hospital plant effluents proving bacteriophages as a significant pool for ARGs, implying towards their possible role in the spread of AMR [63].

2.5.3 Conjugation

Of all the three methods of HGT, conjugation is a widely investigated and most understood mechanism. It is the process of DNA transfer along a multiple-step process through cell-cell contact using cell surface adhesions or pili from the donor to recipient [68]. It is made possible using the conjugative machinery encoded by genes on replicating plasmids or using integrative elements in the chromosome. In many cases, ARDs are related to conjugative elements like transposons or plasmids which can also be transferred through transduction or transformation, but conjugation is considered the most likely mechanism due to its efficiency in entering host cell and enhanced protection from environmental surroundings [69]. Conjugation of mobile genetic elements giving out AMR is noticed in many ecosystems that range from transmission between bacteria present in the soil, insect, and water environments to different healthcare and food-related pathogens [5]. For instance, transfer of conjugative transposons and plasmids Tn916 family occurs among related bacteria and over long taxonomic distances, indicating the contribution of this mechanism in disseminating ARGs between various reservoirs by wide host range of MGEs [70]. The antibiotic resistance plasmid spread in human pathogen shows that when the resistance gene becomes established in thriving plasmid, it becomes capable of spreading rapidly across different strains, genera, or species. For example, blaCTX-M ESβL genes have been disseminated to a wide range of broad and narrow host plasmids in the *Enterobactriacecae* and to another human pathogen [57].

2.6 Conclusion and Future Perspectives

The generation and transfer of new gene factors and resistance genes is a continuous process. The extraordinary growth in antibiotic levels in the environment caused by humans, animals, and microorganisms disrupts the equilibrium balance between antimicrobials and bacteria. It has a wide-ranging impact on numerous communities, resulting in an intangible risk to healthcare because resistance spreads swiftly all over the world. Despite the absence of antibiotic selection, the resistance genes will not be uprooted. As a result, there is a need to focus on the drivers of emerging transmission and get a better understanding of molecular mechanisms in order to reduce the development of antimicrobial resistance, improve AMR mitigation, and gain a deeper understanding of the hazards associated. The natural equilibrium of antimicrobials and microbes has been disrupted due to an increase in antibiotics in use. This has led to the dissemination of these resistance genes. The environment, human and animal guts, and commensals serve as ARG reservoirs, disseminating them through transformation, transduction, and conjugation. Further work to understand the extent of dissemination will help strengthen antibiotic stewardship and will be required for the evidence-based policymaking against AMR. Elucidating the extent of this dissemination is pertinent in the war against antimicrobial resistance.

Bibliography

[1] Ramanan Laxminarayan, Adriano Duse, Chand Wattal, Anita KM Zaidi, Heiman FL Wertheim, Nithima Sumpradit, Erika Vlieghe, Gabriel Levy Hara, Ian M Gould, Herman Goossens, et al. Antibiotic resistance—the need for global solutions. *The Lancet infectious diseases*, 13(12):1057–1098, 2013.

[2] Glenn S Tillotson and Stephen H Zinner. Burden of antimicrobial resistance in an era of decreasing susceptibility. *Expert review of anti-infective therapy*, 15(7):663–676, 2017.

[3] Thomas P Van Boeckel, Charles Brower, Marius Gilbert, Bryan T Grenfell, Simon A Levin, Timothy P Robinson, Aude Teillant, and Ramanan Laxminarayan. Global trends in antimicrobial use in food animals. *Proceedings of the National Academy of Sciences*, 112(18):5649–5654, 2015.

[4] Daniela Pires, Marlieke Elizabeth Adriana de Kraker, Ermira Tartari, Mohamed Abbas, and Didier Pittet. 'fight antibiotic resistance—it's in your hands': call from the world health organization for 5th may 2017. *Clinical Infectious Diseases*, 64(12):1780–1783, 2017.

[5] Julian Davies and Dorothy Davies. Origins and evolution of antibiotic resistance. *Microbiology and molecular biology reviews*, 74(3):417–433, 2010.

[6] Yvonne Pfeifer, Angela Cullik, and Wolfgang Witte. Resistance to cephalosporins and carbapenems in gram-negative bacterial pathogens. *International journal of medical microbiology*, 300(6):371–379, 2010.

[7] Styliani Karanika, Theodoros Karantanos, Marios Arvanitis,

Christos Grigoras, and Eleftherios Mylonakis. Fecal colonization with extended-spectrum beta-lactamase–producing enterobacteriaceae and risk factors among healthy individuals: a systematic review and metaanalysis. *Reviews of Infectious Diseases*, 63(3):310–318, 2016.

[8] Timothy R Walsh, Janis Weeks, David M Livermore, and Mark A Toleman. Dissemination of ndm-1 positive bacteria in the new delhi environment and its implications for human health: an environmental point prevalence study. *The Lancet infectious diseases*, 11(5):355–362, 2011.

[9] Erik Gullberg, Lisa M Albrecht, Christoffer Karlsson, Linus Sandegren, and Dan I Andersson. Selection of a multidrug resistance plasmid by sublethal levels of antibiotics and heavy metals. *MBio*, 5(5):e01918–14, 2014.

[10] SS Chung, JS Zheng, SR Burket, and BW Brooks. Select antibiotics in leachate from closed and active landfills exceed thresholds for antibiotic resistance development. *Environment International*, 115:89–96, 2018.

[11] Rustam I Aminov. A brief history of the antibiotic era: lessons learned and challenges for the future. *Frontiers in microbiology*, 1:134, 2010.

[12] Alison H Holmes, Luke SP Moore, Arnfinn Sundsfjord, Martin Steinbakk, Sadie Regmi, Abhilasha Karkey, Philippe J Guerin, and Laura JV Piddock. Understanding the mechanisms and drivers of antimicrobial resistance. *The Lancet*, 387(10014):176–187, 2016.

[13] D Maxwell Bruce and A Martin Mortimer. Selection for herbicide resistance. In *Herbicide Resistance in Plants*, pages 1–26. CRC Press, 2018.

[14] Pedro Fernandes, Bruno Sommer Ferreira, and Joaquim Manuel Sampaio Cabral. Solvent tolerance in bacteria: role of efflux pumps and cross-resistance with antibiotics. *International journal of antimicrobial agents*, 22(3):211–216, 2003.

[15] Jens Kallmeyer, Robert Pockalny, Rishi Ram Adhikari, David C Smith, and Steven D'Hondt. Global distribution of microbial abundance and biomass in subseafloor sediment. *Proceedings of the National Academy of Sciences*, 109(40):16213–16216, 2012.

[16] Elizabeth MH Wellington, Alistair BA Boxall, Paul Cross, Edward J Feil, William H Gaze, Peter M Hawkey, Ashley S Johnson-Rollings, Davey L Jones, Nicholas M Lee, Wilfred Otten, et al. The role of the natural environment in the emergence of antibiotic resistance in gram-negative bacteria. *The Lancet infectious diseases*, 13(2):155–165, 2013.

[17] Xiuling Ji, Qunhui Shen, Fang Liu, Jing Ma, Gang Xu, Yuanlong Wang, and Minghong Wu. Antibiotic resistance gene abundances associated with antibiotics and heavy metals in animal manures and agricultural soils adjacent to feedlots in shanghai; china. *Journal of Hazardous Materials*, 235:178–185, 2012.

[18] Mark A Webber, Rebekah N Whitehead, Manuella Mount, Nick J Loman, Mark J Pallen, and Laura JV Piddock. Parallel evolutionary pathways to antibiotic resistance selected by biocide exposure. *Journal of antimicrobial chemotherapy*, 70(8):2241–2248, 2015.

[19] Andrew D Wales and Robert H Davies. Co-selection of resistance to antibiotics, biocides and heavy metals, and its relevance to foodborne pathogens. *Antibiotics*, 4(4):567–604, 2015.

[20] Brad Spellberg, Gail R Hansen, Avinash Kar, Carmen D Cordova, Lance B Price, and James R Johnson. Antibiotic resistance in humans and animals. *NAM Perspectives*, 2016.

[21] Jan AJW Kluytmans, Ilse TMA Overdevest, Ina Willemsen, Marjolein FQ Kluytmans-Van Den Bergh, Kim Van Der Zwaluw, Max Heck, Martine Rijnsburger, Christina MJE Vandenbroucke-Grauls, Paul HM Savelkoul, Brian D Johnston, et al. Extended-spectrum β-lactamase–producing es-

cherichia coli from retail chicken meat and humans: comparison of strains, plasmids, resistance genes, and virulence factors. *Clinical Infectious Diseases*, 56(4):478–487, 2013.

[22] Muhammad Sohail, Abid Rashid, Bilal Aslam, Muhammad Waseem, Muhammad Shahid, Muhammad Akram, Mohsin Khurshid, and Muhammad Hidayat Rasool. Antimicrobial susceptibility of acinetobacter clinical isolates and emerging antibiogram trends for nosocomial infection management. *Revista da Sociedade Brasileira de Medicina Tropical*, 49:300–304, 2016.

[23] Sandra R Arnold and Sharon E Straus. Interventions to improve antibiotic prescribing practices in ambulatory care. *Cochrane database of systematic reviews*, (4), 2005.

[24] DA Lewis. The role of core groups in the emergence and dissemination of antimicrobial-resistant n gonorrhoeae. *Sexually transmitted infections*, 89(Suppl 4):iv47–iv51, 2013.

[25] Claire P Thomas, Luke SP Moore, Nazik Elamin, Michel Doumith, Jiancheng Zhang, Sunil Maharjan, Marina Warner, Claire Perry, Jane F Turton, Clare Johnstone, et al. Early (2008–2010) hospital outbreak of klebsiella pneumoniae producing oxa-48 carbapenemase in the uk. *International journal of antimicrobial agents*, 42(6):531–536, 2013.

[26] Charu Kothari, Rajni Gaind, Laishram Chandreshwor Singh, Anju Sinha, Vidya Kumari, Sugandha Arya, Harish Chellani, Sunita Saxena, and Monorama Deb. Community acquisition of β-lactamase producing enterobacteriaceae in neonatal gut. *BMC microbiology*, 13(1):1–6, 2013.

[27] Yonas I Tekle, Kaare M Nielsen, Jingzhou Liu, Melinda M Pettigrew, Lauren A Meyers, Alison P Galvani, and Jeffrey P Townsend. Controlling antimicrobial resistance through targeted, vaccine-induced replacement of strains. *PLoS One*, 7(12):e50688, 2012.

[28] Takele Beyene. Veterinary drug residues in food-animal products: its risk factors and potential effects on public health. *J Vet Sci Technol*, 7(1):1–7, 2016.

[29] Bjorn JA Berendsen, Robin S Wegh, Joost Memelink, Tina Zuidema, and Linda AM Stolker. The analysis of animal faeces as a tool to monitor antibiotic usage. *Talanta*, 132:258–268, 2015.

[30] Tara C Smith, Wondwossen A Gebreyes, Melanie J Abley, Abby L Harper, Brett M Forshey, Michael J Male, H Wayne Martin, Bayleyegn Z Molla, Srinand Sreevatsan, Siddhartha Thakur, et al. Methicillin-resistant staphylococcus aureus in pigs and farm workers on conventional and antibiotic-free swine farms in the usa. *PLoS One*, 8(5):e63704, 2013.

[31] C Hölzel and J Bauer. Salmonella spp. in bavarian liquid pig manure: Occurrence and relevance for the distribution of antibiotic resistance. *Zoonoses and public health*, 55(3):133–138, 2008.

[32] ES Anderson, MJ Lewis, et al. Drug resistance and its transfer in salmonella typhimurium. *Nature*, 206:579–83, 1965.

[33] Hilde Kruse and H Sørum. Transfer of multiple drug resistance plasmids between bacteria of diverse origins in natural microenvironments. *Applied and environmental microbiology*, 60(11):4015–4021, 1994.

[34] Jose Luis Martinez. Environmental pollution by antibiotics and by antibiotic resistance determinants. *Environmental pollution*, 157(11):2893–2902, 2009.

[35] Gong Cheng, Yongfei Hu, Yeshi Yin, Xi Yang, Chunsheng Xiang, Baohong Wang, Yanfei Chen, Fengling Yang, Fang Lei, Na Wu, et al. Functional screening of antibiotic resistance genes from human gut microbiota reveals a novel gene fusion. *FEMS microbiology letters*, 336(1):11–16, 2012.

[36] L Rizzo, C Manaia, C Merlin, T Schwartz, C Dagot, MC Ploy, I Michael, and Despo Fatta-Kassinos. Urban wastewater

treatment plants as hotspots for antibiotic resistant bacteria and genes spread into the environment: a review. *Science of the total environment*, 447:345–360, 2013.

[37] Marit G Tellevik, Bjørn Blomberg, Øyvind Kommedal, Samuel Y Maselle, Nina Langeland, and Sabrina J Moyo. High prevalence of faecal carriage of esbl-producing enterobacteriaceae among children in dar es salaam, tanzania. *PloS one*, 11(12):e0168024, 2016.

[38] Genevieve A Kahrilas, Jens Blotevogel, Philip S Stewart, and Thomas Borch. Biocides in hydraulic fracturing fluids: a critical review of their usage, mobility, degradation, and toxicity. *Environmental science & technology*, 49(1):16–32, 2015.

[39] AD Russell. Biocide use and antibiotic resistance: the relevance of laboratory findings to clinical and environmental situations. *The Lancet infectious diseases*, 3(12):794–803, 2003.

[40] Sylvie Buffet-Bataillon, Pierre Tattevin, Jean-Yves Maillard, Martine Bonnaure-Mallet, and Anne Jolivet-Gougeon. Efflux pump induction by quaternary ammonium compounds and fluoroquinolone resistance in bacteria. *Future microbiology*, 11(1):81–92, 2016.

[41] Chandan Pal, Johan Bengtsson-Palme, Erik Kristiansson, and DG Joakim Larsson. Co-occurrence of resistance genes to antibiotics, biocides and metals reveals novel insights into their co-selection potential. *BMC genomics*, 16(1):1–14, 2015.

[42] Michalis Karvelas, Athanasios Katsoyiannis, and Constantini Samara. Occurrence and fate of heavy metals in the wastewater treatment process. *Chemosphere*, 53(10):1201–1210, 2003.

[43] Henriette Selck, Richard D Handy, Teresa F Fernandes, Stephen J Klaine, and Elijah J Petersen. Nanomaterials in the aquatic environment: A european union–united states perspective on the status of ecotoxicity testing, research priorities, and challenges ahead. *Environmental Toxicology and Chemistry*, 35(5):1055–1067, 2016.

[44] Craig Baker-Austin, Meredith S Wright, Ramunas Stepanauskas, and JV McArthur. Co-selection of antibiotic and metal resistance. *Trends in microbiology*, 14(4):176–182, 2006.

[45] Charles W Knapp, Seánín M McCluskey, Brajesh K Singh, Colin D Campbell, Gordon Hudson, and David W Graham. Antibiotic resistance gene abundances correlate with metal and geochemical conditions in archived scottish soils. *PloS one*, 6(11):e27300, 2011.

[46] Masoud M Ardestani, Nico M van Straalen, and Cornelis AM van Gestel. Biotic ligand modeling approach: synthesis of the effect of major cations on the toxicity of metals to soil and aquatic organisms. *Environmental Toxicology and Chemistry*, 34(10):2194–2204, 2015.

[47] Andrea Di Cesare, Ester M Eckert, Silvia D'Urso, Roberto Bertoni, David C Gillan, Ruddy Wattiez, and Gianluca Corno. Co-occurrence of integrase 1, antibiotic and heavy metal resistance genes in municipal wastewater treatment plants. *Water Research*, 94:208–214, 2016.

[48] Paul S Morley, Michael D Apley, Thomas E Besser, Derek P Burney, Paula J Fedorka-Cray, Mark G Papich, Josie L Traub-Dargatz, and J Scott Weese. Antimicrobial drug use in veterinary medicine. *Journal of veterinary internal medicine*, 19(4):617–629, 2005.

[49] Wanda C Reygaert. An overview of the antimicrobial resistance mechanisms of bacteria. *AIMS microbiology*, 4(3):482, 2018.

[50] Claus Sibold, Jørgen Henrichsen, Andrea König, Christiane Martin, Lynda Chalkley, and Regine Hakenbeck. Mosaic pbpx genes of major clones of penicillin-resistant streptococcus pneumoniae have evolved from pbpx genes of a penicillin-sensitive streptococcus oralis. *Molecular microbiology*, 12(6):1013–1023, 1994.

[51] Maite Muniesa, Aurora García, Elisenda Miró, Beatriz Mirelis, Guillem Prats, Juan Jofre, and Ferran Navarro. Bacteriophages and diffusion of β-lactamase genes. *Emerging infectious diseases*, 10(6):1134, 2004.

[52] Gerard D Wright. Antibiotic resistance in the environment: a link to the clinic? *Current opinion in microbiology*, 13(5):589–594, 2010.

[53] Gabriel G Perron, Lyle Whyte, Peter J Turnbaugh, Jacqueline Goordial, William P Hanage, Gautam Dantas, and Michael M Desai. Functional characterization of bacteria isolated from ancient arctic soil exposes diverse resistance mechanisms to modern antibiotics. *PloS one*, 10(3):e0069533, 2015.

[54] Katarzyna A Kazimierczak, Karen P Scott, Denise Kelly, and Rustam I Aminov. Tetracycline resistome of the organic pig gut. *Applied and environmental microbiology*, 75(6):1717–1722, 2009.

[55] Sara Domingues, Klaus Harms, W Florian Fricke, Pål J Johnsen, Gabriela J Da Silva, and Kaare Magne Nielsen. Natural transformation facilitates transfer of transposons, integrons and gene cassettes between bacterial species. 2012.

[56] Larissa C Parsley, Erin J Consuegra, Kavita S Kakirde, Andrew M Land, Willie F Harper Jr, and Mark R Liles. Identification of diverse antimicrobial resistance determinants carried on bacterial, plasmid, or viral metagenomes from an activated sludge microbial assemblage. *Applied and environmental microbiology*, 76(11):3753–3757, 2010.

[57] Peter M Hawkey and Annie M Jones. The changing epidemiology of resistance. *Journal of antimicrobial chemotherapy*, 64(suppl_1):i3–i10, 2009.

[58] Christopher M Thomas and Kaare M Nielsen. Mechanisms of, and barriers to, horizontal gene transfer between bacteria. *Nature reviews microbiology*, 3(9):711–721, 2005.

[59] Philip F Sparling. Genetic transformation of neisseria gonorrhoeae to streptomycin resistance. *Journal of bacteriology*, 92(5):1364–1371, 1966.

[60] Brian G Spratt, Qian-Yun Zhang, Dennis M Jones, Agnes Hutchison, James A Brannigan, and Christopher G Dowson. Recruitment of a penicillin-binding protein gene from neisseria flavescens during the emergence of penicillin resistance in neisseria meningitidis. *Proceedings of the National Academy of Sciences*, 86(22):8988–8992, 1989.

[61] Luz Balsalobre, María José Ferrándiz, Josefina Linares, Fe Tubau, and Adela G de la Campa. Viridans group streptococci are donors in horizontal transfer of topoisomerase iv genes to streptococcus pneumoniae. *Antimicrobial Agents and Chemotherapy*, 47(7):2072–2081, 2003.

[62] John Davison. Genetic exchange between bacteria in the environment. *Plasmid*, 42(2):73–91, 1999.

[63] Sheetal R Modi, Henry H Lee, Catherine S Spina, and James J Collins. Antibiotic treatment expands the resistance reservoir and ecological network of the phage metagenome. *Nature*, 499(7457):219–222, 2013.

[64] Sylvia L Hyder and Murray M Streitfeld. Transfer of erythromycin resistance from clinically isolated lysogenic strains of streptococcus pyogenes via their endogenous phage. *Journal of Infectious Diseases*, 138(3):281–286, 1978.

[65] Horst Schmieger and Petra Schicklmaier. Transduction of multiple drug resistance of salmonella enterica serovar typhimurium dt104. *FEMS microbiology letters*, 170(1):251–256, 1999.

[66] Marian Varga, Lucie Kuntová, Roman Pantček, Ivana Mašlaňová, Vladislava Ržičková, and Jiří Doškař. Efficient transfer of antibiotic resistance plasmids by transduction within methicillin-resistant staphylococcus aureus usa300 clone. *FEMS microbiology letters*, 332(2):146–152, 2012.

[67] Jean Marc Rolain, Laura Fancello, Christelle Desnues, and Didier Raoult. Bacteriophages as vehicles of the resistome in cystic fibrosis. *Journal of antimicrobial chemotherapy*, 66(11):2444–2447, 2011.

[68] Chris Smillie, M Pilar Garcillán-Barcia, M Victoria Francia, Eduardo PC Rocha, and Fernando de la Cruz. Mobility of plasmids. *Microbiology and Molecular Biology Reviews*, 74(3):434–452, 2010.

[69] Anders Norman, Lars H Hansen, and Søren J Sørensen. Conjugative plasmids: vessels of the communal gene pool. *Philosophical Transactions of the Royal Society B: Biological Sciences*, 364(1527):2275–2289, 2009.

[70] Adam P Roberts and Peter Mullany. A modular master on the move: the tn916 family of mobile genetic elements. *Trends in microbiology*, 17(6):251–258, 2009.

3

Alternatives to Combat AMR: Hunt for Novel Antimicrobials

Amrita Bhagwat

Symbiosis School of Biological Sciences, Symbiosis International (Deemed University) Pune, 412115, India

Tiyasa Haldar

Symbiosis School of Biological Sciences, Symbiosis International (Deemed University) Pune, 412115, India

Sunil D. Saroj

Symbiosis School of Biological Sciences, Symbiosis International (Deemed University) Pune, 412115, India

CONTENTS

3.1 Introduction

In our daily life microorganisms play crucial role, human is a harbour for numerous microorganisms. The interaction between microorganisms and humans may be both harmful and beneficial, since some cause illnesses in humans while others provide useful products for humans through biotechnology. Microorganisms, such as small virus particles, unicellular bacteria, protozoa, and fungus, are the causative agents for many infectious diseases in humans. Antimicrobial agents, development is a significant step forward in the fight against pathogenic microorganisms. Medical professionals hailed this development as a breakthrough in the fight against pathogen-caused diseases. Nonetheless, disease-causing microscopic devils plagued the pre-antibiotic age, and human history is littered with evidence of terrible illnesses like smallpox, chicken pox, polio, typhoid, cholera, diphtheria, TB, and so on. When a novel antimicrobial is introduced, drug resistance develops quickly, and this is true for antibiotics, antivirals, and antifungal medicines. Antimicrobial therapy puts the organism under selection pressure, for which microorganism develop different resistance mechanism. Due to the extensive and inappropriate use of different broad range of antibiotics without having proper knowledge of antimicrobial susceptibility patterns has led to the development of antimicrobial resistance (AMR) in bacteria. The rate of bacterial resistance towards common antimicrobial therapies and development of multidrug resistant bacteria are exponentially increasing. Every year, 700,000 people die from illnesses that are resistant to treatment. The British Government in their reports have predicted that if the present rate of rise continues, 10 million people will die as a result of AMR by 2050. Thus, understanding the mechanisms underlying AMR, will open up new ways to target and interfere with vital virulence-associated traits of bacteria [1]. Depending on the mode of actions and target sites there are different classes of antibiotics such as, β-lactams, carbapenems, cephalosporins, glycopeptides that are known to inhibit the growth of bacteria by interrupting transpeptidation process of cell wall synthesis [2], while sulphonamides and trimethoprim target folate synthesis [3].

Quinolones are reported to inhibit nucleic acid synthesis by trapping the gyrase or topoisomerase IV [4]. Macrolides, chloramphenicol, aminoglycosides and tetracycline interact with the RNA subunits and inhibits the protein synthesis [5].

Recently World Health Organization (WHO) enlisted most frequently isolated priority pathogens that are resistant to different antibiotics causing severe diseases into three major groups based on priority: critical, high, and medium priority., amongst which *Acinetobacter baumannii* (carbapenem resistant), *Pseudomonas aeruginosa* (carbapenem resistant), *Enterobacteriaceae* are mentioned as critical pathogens, *Enterococcus faecium* (vancomycin-resistant), *Staphylococcus aureus* (methicillin, vancomycin -resistant), *Helicobacter pylori* (clarithromycin-resistant), *Campylobacter* spp. (fluoroquinolone-resistant) *Salmonella* (fluoroquinolone-resistant), *Neisseria gonorrhoeae* (cephalosporin and fluoroquinolone-resistant) as high risk priority pathogens. *Haemophilus influenzae* (ampicillin-resistant), *Shigella* spp.(fluoroquinolone-resistant), *Streptococcus pneumoniae* (no more susceptible in Penicillin) are under moderate risk pathogens. According to current report by WHO, approximately 700,000 people worldwide die because of antimicrobial resistance disease whereas 230,000 die due to multidrug resistance, respiratory infection, sexually transmitted disease, and urinary tract infections. So, clearly the emergence of AMR is one of the major health concerns. On the other side, conventional antimicrobials are no more useful to destroy superbugs because of their high adaptability to survive in antimicrobial stress. This creates an urge to discover new antimicrobials, but discovery of new antimicrobial agents is challenging due to the rapid increase of AMR. In different parts of the world, disease resistance differs. As a result, local resistance patterns must be understood in order to employ antimicrobials effectively. Following the strong growth in availability to antiretroviral drugs in recent years, resistance is becoming a problem for HIV therapy. Several Gram-negative bacteria have developed new resistance mechanisms, such as the metallobeta-lactamase NDM1. This can render potent antibiotics, which are frequently employed as a final line of defense against multiresistant bacteria, ineffective.

We know bacterial pathogenesis depends on many factors that are related to host as well as host microbiome. Host environment is dynamic and bacteria needs to acclimatize to these changing environments in order to establish successful infection. Bacteria are capable of colonizing and surviving as biofilms in varying conditions that it encounters in host. The bacterial biofilm plays a major role in various diseases caused by pathogenic organisms as it acts as a defensive barrier against treatment therapies used to eradicate biofilms. Bacteria that grow inside quorum sensing (QS) mediated biofilms cause the majority of infectious illnesses. Bioactive compounds generated by prokaryotes and eukaryotes have been identified resulting in destruction of biofilm [6]. These molecules function as quorum sensing quenchers or inhibitors (QSIs), while the phenomenon is termed to be quorum quenching (QQ). There are various synthetic as well as natural compounds that act as inhibitors of quorum sensing. It has been claimed that the most efficient prokaryotes for producing QSI are ones that are typically considered safe. Certain legumes and traditional medicinal herbs are expected to behave as QSIs in eukaryotes. Such discoveries are anticipated to lead to more effective therapies using considerably lower medication dosages, particularly antibiotics, than are now necessary [7].

3.2 Antimicrobial Resistance

The chemical substances produced by different microorganisms which have the capacity to inhibit the growth or kill the bacteria and other microorganisms are known as antimicrobial agents. The ability of a disease-causing microorganisms like that of viruses, bacteria, and parasites to survive in presences of antimicrobials is referred to as antimicrobial resistance. AMR can be easily originated from environment. In fact, the producers of such antimicrobial substances can also carry forward different AMR genes to the environment from where bacteria and other microorganisms can acquire those via horizontal gene transfer. Prolonged, random use of multiple antibiotics for different diseases caused by pathogens has made the pathogens develop different resistance mechanisms leading to the infection becoming difficult to cure or incurable. Epidemiological data on common bacterial pathogens and their antibiotic resistance will give insight into appropriate treatment options, allowing hospitals to develop an effective infection prevention and control programme [8]. Lack of understanding about how to use antibiotics, poor dosages, and no early detection of infectious disease and their antimicrobial sensitivity pattern in patients with bacteremia and other severe diseases are all factors that contribute to the emergence of antimicrobial resistance. There is rapid increase in antimicrobial resistance as compared to the infection control procedures, resulting in quick spread of resistant organism to other patients and the environment. Previously the AMR pathogens such as *Acinetobacter* sp. were restricted to the clinical environment, but recently numerous studies confirm that it is also found in farm and food products [9]. Bacteria and other pathogens frequently get exposed to a wide range of different stresses from the outer environment as well as their host environment. Owing to combat all the stresses like starvation, temperature, and most importantly, exposure of multiple antibiotics at different concentrations make the bacteria develop different resistant mechanisms. Bacteria and other pathogens follow three different pathways to acquire resistance against antimicrobial agents which are intrinsic, extrinsic, and adaptive. When a microorganism species has innate capacity to resist certain antibiotics, i.e., natural insensitivity to antibiotics due to its own functional characteristic and other inherent factors, then it possesses intrinsic resistance. For example, anaerobe bacteria do not consist of an oxidative metabolism system so they cannot uptake aminoglycoside antibiotics which made them become intrinsic resistant against aminoglycosides [10]. The reasons behind acquiring resistance to the antibiotics could be 1) production of enzymes that inactivate the antibiotic, 2) modification of target site, so that there is no affinity between antibiotic and the target site and 3) extrusion of antibiotic by activated exporter system such as different multidrug efflux pumps [11]. Sometimes the outer membrane impermeability of the bacteria also prevents antibiotics transporting into the cell as an example in *Pseudomonas* sp. and *Acinetobacter* sp. possess such membrane impermeability leading to MDR [12]. As previously mentioned, that environment is the most important source for bacteria and other microorganisms to develop antibiotic resistance via horizontal gene transfer (HGT) and mutation. As bacteria have shorter generation time with the help of transformation, conjugation they have the potential to exchange of genetic materials within species which give them the advantages to adapt in the different niche with respect to response for different stresses like antibiotics [13]. Bacteria can acquire AMR through a variety of mutations in addition to gene transfer. For example, mutation in DNA topoisomerase which acts as target for fluoroquinolone will lead to severe resistance against different bacteria [14]. Eventually overall adaptation of bacteria with respect to stress response is the most important factors for emerging adaptive antimicrobial resistance. Antibiotic inactivation is an adaptive process of AMR in which bacteria develop the capacity to inactivate antibiotic molecules [15]. Not only via inactivating, bacteria can also modify the antibiotic target site or bypass the target to become antibiotic insensitive [16]. On the other hand there are different bacterial species that have the ability to form biofilm, i.e., polysaccharide structure around the bacterial colony, most of the antibiotics cannot penetrate and reach the bacteria or very small concentration (sub-inhibitory) of antibiotics gets in touch with bacteria so they remain insensitive to the antibiotics such as *Neisseria* sp., *Listeria monocytogenes*, *Salmonella* sp. [17]. Besides forming biofilm and spores, many bacteria undergo to the formation of the viable but non-culturable state (VBNC) where to survive from stresses they show dormancy but after relaxation of adversity they regain their normal behavior [18]. One more aspect of the acquisition of antibiotic resistance genes is through bacterial conjugative plasmids with the help of transposition and site-specific recombination, because bacterial plasmid possesses different array of AMR genes which can be easily transferred within different species [19]. The chapter includes an elucidatory table (Table 3.1) that gives brief descriptions regarding the class, mode of action, and resistance mechanism of various antibiotics.

TABLE 3.1
Different antibiotics of different classes, their mode of action and the resistance mechanism of bacteria against it

Class of antibiotic	Name of antibiotics	Mode of actions	Resistance mechanism of bacteria
β-Lactams	Penicillin	Inhibition of cell wall synthesis	Modification in penicillin binding protein (PBP), production of penicillin degrading enzymes (β-Lactamase) (R. Williamson, Collatz, and Gutmann 1986)
	Ampicillin	Inhibition of cell wall synthesis	Gutmann 1986).
	Amoxicillin	Inhibition of cell wall synthesis	Changes in membrane permeability, enzymatic degradation of antibiotic (R. Williamson, Collatz, and Gutmann 1986).
	Cephalosporin	Inhibition of cell wall synthesis	Production of β-Lactamase (R. Williamson, Collatz, and Gutmann 1986).
	Methicillin	Inhibition of cell wall synthesis	Modification in PBP, PBP2a (R. Williamson, Collatz, and Gutmann 1986)
	Cephamycin	Inhibition of cell wall synthesis (inhibit mucopeptide)	Production of β-Lactamase (R. Williamson, Collatz, and Gutmann 1986)
Sulphonamides	Gantrisin	Inhibit folate synthesis (Inhibition to utilization of para-aminobenzoic acid PABA).	Horizontal transfer for folP gene and its parts which encodes for dihropteroate synthase enzyme (Henry 1943)
	Trimethoprim	Inhibit folate synthesis.	reductase (Henry 1943).
Glycopeptides	Vancomycin	Inhibition of cell wall synthesis	Changing the membrane permeability (Davies and Davies 2010).
	Carbomycin	Inhibition of cell wall synthesis	Changing membrane permeability.
Aminoglycosides	Streptomycin	Inhibition of protein synthesis (30S sub-unit)	Enzymatic modification, degradation of antibiotic (G. M. Williamson 1957).
	Tobramycin	Inhibition of protein synthesis (30S sub-unit)	Enzymatic modification, degradation of antibiotic.
	Neomycin	Inhibition of protein synthesis	Enzymatic modification, degradation of antibiotic (Kile, Rockwell, and Schwarz 1952).
	Gentamycin	Inhibition of protein synthesis (30S sub-unit)	Enzymatic modification, degradation of antibiotic (Krause et al. 2016).
Rifampin	Rifampicin	Inhibition of DNA dependent RNA synthesis	Failure of chemoprophylaxis and change in membrane permeability (Sensi 1983).
Fluoroquinolone	Ciprofloxacin	Inhibition of nucleic acid synthesis (DNA gyrase)	Presence of efflux-pump mechanism (Blondeau 2004)
Macrolides	Azithromycin	Inhibition of protein synthesis (50S sub-unit)	target modification consists of post-transcriptional methylation of an adenine residue in 23S rRNA. They have also active efflux mechanism (Patel PH 2022).
	Erythromycin	Inhibition of protein synthesis (50S sub-unit)	Drug efflux-pump mechanism(Patel PH 2022).
Tetracycline	Tetracycline	Inhibition of protein synthesis	Using efflux-pump to elute out the antibiotic (Chopra and Roberts 2001).
	Minocycline	Inhibition of protein synthesis	Using efflux-pump to elute out the antibiotic.

3.2.1 National Antimicrobial Resistance Policy, India

India implemented an antimicrobial resistance (AMR) policy in 2011. Its goal is to understand the factors that influence AMR's emergence and spread. The antimicrobial programme has been developed as part of the policy, which stimulates the development of new effective antimicrobials while simultaneously rationalizing the use of present antibiotics. In addition, building an AMR surveillance system that can help prevent infection by implementing control methods, educating, training, and encouraging all stakeholders in the judicious use of antimicrobials were some of the policy priorities. According to the WHO, only around half of all countries have basic protocols in place to guarantee that pharmaceuticals are administered properly [20].

3.3 Current status of antibiotics

Since the discovery of penicillin different antibiotics are developed to combat several pathogenic diseases. Diverse pathogenic organisms have developed different resistance mechanisms to survive, and antibiotic stress is one of the most common exposures for different microorganisms in the host body since different antimicrobials are used on a daily basis in both animals and humans. Now AMR is one of the most important concerns for world. Because of the availability of more accessible and inexpensive medications in the market, AMR for HIV, typhoid fever, malaria, and other diseases are on the rise in low- and middle-income nations. According to WHO the use of first line antibiotics necessary in daily life has increased to 9 percent worldwide and 165 percent in the low- and middle-income countries within 15 years. From the data of The Centre for Disease Dynamics, Economics and Policy different bacteria possess different rates of resistance for an antibiotic. Resistance to commonly used medicines such as ciprofloxacin, which is used to treat urinary tract infections, has climbed 4.8 percent to 82 percent in *E. coli*, according to a recent WHO study from 66 countries. In 2019, the six most common pathogens for deaths associated with resistance (*Escherichia coli, Staphylococcus aureus, Klebsiella pneumoniae, Streptococcus pneumoniae, Acinetobacter baumannii,* and *Pseudomonas aeruginosa*) were responsible for 929 000 (660 000–1 270 000) deaths attributable to AMR and 357 million (262–478). Eventually there is a huge impact in the use of antibiotics for the current COVID-19 pandemic. Because of misinformation about the drugs thought to be beneficial for COVID-19 there are of 71 percent patients who had already taken one dose of antibiotics despite having just 3.6 percent of co-infection cases [21]). Although there are different vaccines available the problem arises due to the lack of vaccine coverage, proper sanitization, and hygiene practice. For that reason, people have to depend on the antibiotics for any disease control. So, in this situation there is, need for alternative treatment of antibiotics. There are different preventive measures for antimicrobial resistance already taken by WHO and other global organization such as "one health approach." But all these measures remain as a concept because it is not implemented in a proper way. In case of disease caused by ESKAPE (*Enterococcus faecium, Staphylococcus aureus, Klebsiella pneumoniae, Acinetobacter baumannii, Pseudomonas aeruginosa, Enterobacter* sp.) which are highly antimicrobial resistance and listed as the most critical priority pathogen according to WHO, are no more sensitive to the available antibiotics. So, obviously there is a great shortage of new effective antibiotic against these pathogens [22]. According to the report of 2019 about 41 new antibiotics are under development among them, 15 are in phase I, 12 are phase II, and 13 are under phase III clinical trial. It has been found that 18 of these have the ability to act against ESKAPE. From the beginning of the discovery of antibiotics there are 65 different classes of antibiotics that are derived but currently there is research to invent more antibiotics from the conventional antibiotics exist [23]. Though available antibiotics are no more effective the use of different antibiotics is also unavoidable. So, different newly synthesized advanced antimicrobials have been commercially manufactured. Recently an antibiotic named Telavancin has been synthesized which is an advanced derivative of Vancomycin [24]. But with respect to the emerging threat of AMR the necessity to develop alternative strategies is the most important. After 2012, the

GAIN (Generating Antibiotics Incentives Now) Act was enacted by the Food and Drug Administration Safety and Innovation Act in response to the AMR epidemic. Within the years of 2012-2017 there were 12 new drugs approved by Food and Drug Administration (FDA) [25]. With context of *M. tuberculosis* infection, which is a major concern in the underdeveloped countries, it is very difficult to combat this disease due to the lack of available antibacterial drugs. Currently WHO recommended more new antibiotics experimental trials. Not only the new antibiotics, the combination of different antibiotics are also one more effective treatment strategy has been started.

3.4 The need for alternative antibiotics

Over past few decades the conditions of AMR are becoming widespread and clinically significant. Development of MDR organism is a possible sign of heading towards post antibiotic era, which is an alarming sign of threat to human and animal health [26]. Lack of thorough preventive measures, shortage of current effective treatment drugs, a few new antimicrobials in the clinical pipeline have added up to the challenges in the combat of bacterial infections. To deal with the alarming rate of antibiotic resistances, we need to have clearer understanding of bacterial virulence and infection induced molecular pathways [1]. There is an urgent need to investigate alternative treatment options while there are still a few antibiotics left. Before the development of antibiotics, some efforts were made to use bacteriophages as a treatment option, but most of this research stopped soon after the discovery of antibiotics. There are a few potential alternatives like use of quorum sensing inhibitors (QSI)/ Quorum Quencher (QQ) that are used against biofilm-forming organisms [27]. The economic cost, mortality, and morbidity associated with biofilm-forming antibiotic resistant organisms are all important considerations. As a result, biofilm development and antibiotic resistance in an organism reduce the efficacy of existing medications, adding to the healthcare system's constraints. The QSIs or QQ compounds are greatly effective in combination to the antibiotics used in the treatment of biofilm-forming pathogens. Natural resources like plant, fungi, bacteria, and marine bacteria are potent sources of bioactive that exhibit anti-biofilm activity. Bioactive and different synthetic compounds can act as quorum quenchers by either biofilm inhibition or eradication. Use of bacteriophage is another alternative to the antimicrobials. High success rate and no toxic effects in hosts are some major advantages of bacteriophage therapy (Rohde and Kutter 2018). The above-mentioned methods are most accepted replacements to antimicrobials to date. This study will open up new ways to target and interfere with vital virulence-associated traits of bacteria while avoiding the bacterium's evolutionary pressure to develop resistance. The chapter elaborates on the use of bacteriophage and quorum sensing inhibitors derived from vivid sources to combat antimicrobial resistance.

3.5 Bacteriophages

A decade before the discovery of penicillin, use of phage therapy to treat bacterial infections was practiced. In early 1919, pathogen like *Shigella dysenteriae* was treated using bacteriophage.

Phages are non-living organisms consisting of RNA or DNA enclosed in protein capsid. Phages depend on bacteria for their survival as well as reproduction. They are specific to the receptors on the bacterial cell surface, where they bind and inject the genetic material into the host. The genetic material is either integrated into the host genome (temperate phage) or take the control bacterial replication system to produce its own progeny, lysing the host (lytic phage) [28]. For treatment against bacterial infections, phage therapy relies on lytic cycle, where phages obligately kill their host [29]. "Phage cocktails" consisting of multiple phages have proved to be an efficient method in targeting pathogens [30].

Phage therapy is widely implied against a range of clinically important pathogens in different animal models. In comparison to the control group, oral treatment of phage to mice challenged with gut derived sepsis induced by *P. aeruginosa* rescued 66.7 percent

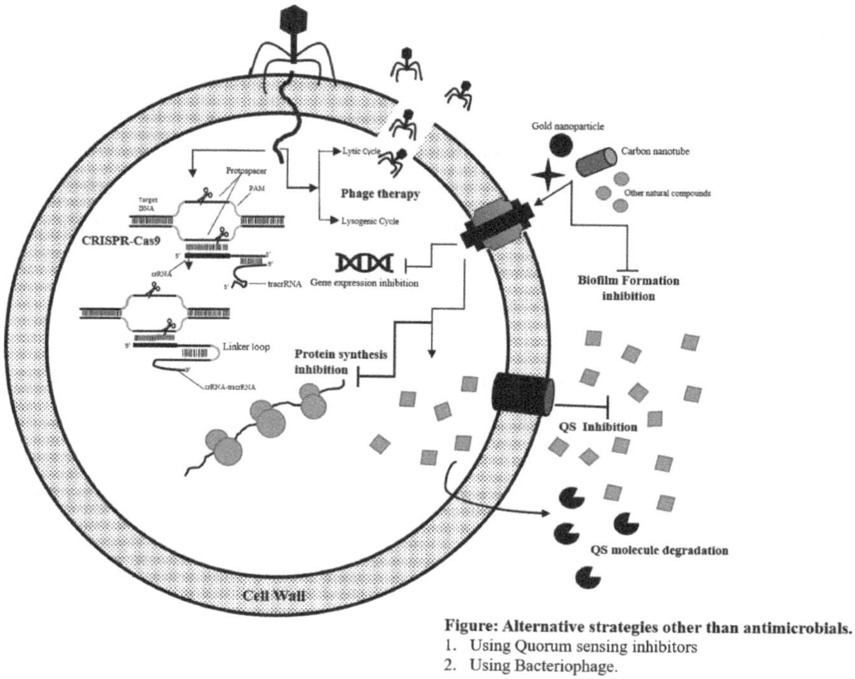

FIGURE 3.1
Use of alternative strategies to combat AMR.

of them. In numerous animal models, *P. aeruginosa* that causes infections of the lungs, skin, and gastrointestinal system was treated by phage cocktail treatment. There is evidence that phage can restore antibiotic sensitivity in bacteria that have become resistant to antibiotics, such as multidrug-resistant *P. aeruginosa* [31]. Human phage therapy studies have been conducted for nearly a century in numerous Eastern European institutes, the most well-known of which are the Eliava Institute of Bacteriophage and the Institute of Immunology and Experimental Therapy in Wroclaw, Poland. *S. aureus, E. coli, Streptococcus* spp., *P. aeruginosa, Proteus* spp., *S. dysenteriae, Salmonella* spp., and *Enterococcus* spp. have all been treated with phage at the Eliava Institute in preclinical and clinical trials [32].

Patients with antibiotic-unresponsive diabetic foot ulcers were treated with S. aureus specific phage, which resulted in adequate healing [30]. A clinical trial in 1938 for phage therapy against bacterial dysentery, targeted *Shigella shiga, Shigella flexneri, E. coli, Proteus* spp., P. *aeruginosa, Salmonella paratyphi* A and B, *Staphylococcus* spp., *Streptococcus* spp. and *Enterococcus* spp. Phage cocktail targeting above pathogens was administered rectally and orally to 219 patients. Amongst 219 patients, 74 percent showed sign of improvement or were completely relieved of the symptoms [33]. To date there are no phage treatment products licensed for human use in EU and the US. But on the other hand, several commercial phage formulations used for biocontrol of bacterial diseases in food sector have been authorized by the FDA as "generally considered as safe." These formulations were used in biocontrol of *Salmonella* spp., *Mycobacterium tuberculosis, Campylobacter* spp., *Listeria monocytogenes*, MRSA, *E. coli* O157:H7, and *Pseudomonas syringae*, among others [34, 33]. Evidence shows that phage biocontrol can improve food safety at a variety of phases in the meat production and processing process, as well as minimise bacterial contamination in fruits, vegetables, and dairy products. These studies into phage biocontrol in food production, as well as recent placebo-controlled human trials demonstrating the safety of oral phage treatment, are gradually filling the phage therapy safety knowledge gap.

With the advent of gene editing tools like that of CRISPR/Cas have added to the efficacy of phage therapy in treatment of bacterial infections. To deliver the CRISPR/Cas system programmed to disrupt antibiotic resistance gene, bioengineered phage is used. These phages might be used to minimise the prevalence and spread of antibiotic resistance genes on hospital surfaces. Bioengineered phages are still in their infancy, but they will definitely deliver numerous beneficial innovations like this one [35].

3.5.1 Advantages and Limitations

One of the major advantages in use of phage therapy is its specificity towards both species and strain. On the other hand, antibiotics have collateral effects on gut commensal microbes, leading to *C. difficile* infection and antibiotic-associated diarrhea [35]. Due to the adverse effect of broad-spectrum antibiotics on gut flora, it can also lead to increased risk of asthma, obesity, and diabetes.

Even though the present understanding of phage treatment's collateral damage is poor, however when compared to antibiotics, phage therapy has been shown to cause less disruption of the gut microbiome successfully eliminating the pathogens [36, 37]. There are reports of various adverse reactions that are associated with the antibiotics like gastrointestinal complications, cardiotoxicity, hepatotoxicity, anaphylaxis, nephrotoxicity etc [38]. The phage treatment, on the other hand, has just lately acquired popularity in western medicine, with little information on its safety accessible. Even while oral phage delivery is thought to be safe, the translocation of phage across the intestinal epithelium, where they then circulate in the blood, is a major concern. Phage translocation downregulates the immune response to indigenous gut microbe antigen, thus benefiting the host. While the benefits of phage therapy certainly exceed the disadvantages in non-immunocompromised patients, the immunological response to phage might indicate the possibility of an unfavorable reaction in immunocompromised individuals, perhaps worsening their condition. In addition to this, another disadvantage associated with the use phage cocktail is induction of "leaky gut" or state of intestinal barrier dysfunction. The potential for phage treatment to impair normal intestinal barrier function might have major consequences for a number of diseases newly related to intestinal barrier dysfunction, including Crohn's disease, inflammatory bowel disease, and type 1 diabetes [36].

Antibiotic therapy is highly effective with planktonic bacteria, such as *V. cholerae* and *Yersinia pestis*, yet is limited in treating biofilm-based bacterial infection. Phages, on the other hand, have enzymes (EPS depolymerase) on the outside of their capsids that destroy extracellular polymeric substances (EPS) and disperse bacterial biofilms, allowing them to reach bacteria embedded in the EPS matrix [39]. High dosages of antibiotics are often necessary to penetrate dense biofilms and notice any suppression of bacterial growth, yet total eradication is not achieved. Once the treatment of antibiotics is stopped, rebuilding of biofilm occurs. Although many antibiotics are typically regarded non-toxic at low dosages, large quantities can cause tissue damage [26]. Biofilms produced by *L. monocytogenes*, *P. aeruginosa*, and *Staphylococ-*

cus epidermidis on the surface of medical equipment have been eradicated using phage treatments. These findings are particularly relevant to the problem of chronic infections induced by implanted medical devices such as catheters, lenses, and prostheses, which frequently produce biofilms [34]. Due to major diversity that exists in bacteriophages, optimizing a phage cocktail becomes complicated as compared to designing a combined antibiotic therapy regimen. The phage cocktail's composition is crucial to phage therapy's effectiveness. Customizing phage therapy to each infection is costly as well as time consuming [40, 34].

3.6 Quorum Sensing Inhibitors

Bacteria use the Quorum Sensing System (QSS) to interact with one another. When bacterial density is low, quorum sensing proves both expensive and ineffectual. A diverse spectrum of microorganisms employ the QSS to defend themselves by creating antibiotics, sporulation, biofilm formation, and even bioluminescence. The association of bacteria with living forms can be symbiotic or it can be pathogenic. The infection causing organisms are the major cause of concern, resulting high mortality and morbidity in humans. Majority of these infectious organisms tend to form biofilms, which has become new target of pharmacological industries. Bacteria generate biofilms as a result of quorum sensing (QS), a cell density-dependent process. Compared to their planktonic counterparts, bacteria found in QS-mediated biofilms have been shown to withstand up to 1000 times greater antibiotic doses [27]. One of major advantage of targeting QS is that it is not directly involved in the growth of the bacteria. Therefore, inhibition of QS does not add to the selective pressure for developing resistance as that of antibiotics. Several QS-inhibitory compounds have been discovered by screening a large number of chemical libraries, both natural and synthetic [7]. These QSI compounds have played crucial role in clearing of infections and reducing mortality in cases of animal models. QS-regulated gene expression in *P. aeruginosa* contributes to biofilm formation and its resistance to antimicrobials. Therefore,

QS system becomes prime target for a novel class of antimicrobial drugs. The QSIs can function by blocking the receptor participating in QS signal in vivo, and successfully attenuating the pathogenesis. *P. aeruginosa* PAO1 shows antibiotic tolerance through the induction of VqsM, which is involved in expression of nfxB (antibiotic resistance regulator). The QSS (Quorum Sensing System) regulator VqsM, leads to transcription of mexC-mexD-oprJ operon that enabled bacteria to tolerate quinolones, tetracycline, and kanamycin [7]. Bacteria that can detect this QS communication have evolved the ability to disrupt it at various stages. Various mechanisms have been documented for a wide spectrum of QSIs: i) inhibiting signal molecule synthesis; (ii) enzymatically degrading signal molecule; (iii) competing with signal molecules for binding to receptor sites; (iv) interfering with signal molecule binding to gene promoters and inhibiting gene expression; and (v) scavenging of AIs by antibodies and macromolecules such as cyclodextrins [41].

3.6.1 Natural compounds as Quorum Quenchers

Plants, animals, fungi, bacteria, and marine sponges produce a variety of chemicals that have anti-biofilm action. Plant extracts are active sources of effective antimicrobial compounds that can combat planktonic mass or biofilms of the pathogen. The discovery of novel antimicrobials will be aided by clinical validation of these bioactives. Thus, adding to the benefits of medical profession with alternative preventative and therapy strategies. The antibiofilm activity is reported in large variety of edible and medicinal plants. Methanolic leaf extract of *Pistacia atlantica* (Anacardiaceae) has four bioactives: rutin, myricetin-3-O-rutinoside, kaempferol-3-O-rutinoside, and isoquercitrin. All the four bioactives extracted from *Pistacia atlantica* can be used against *P.aeruginosa* biofilm infection. *P. aeruginosa* biofilm formation was inhibited by rutin and myricetin 3-O-rutinoside by greater than 90 percent at 0.31 mg/ml. Biofilm formation was inhibited by greater than 90 percent and 70 percent, respectively, by isoquercitin (0.19 mg/ml) and kaempferol-3-O-rutinoside (0.3 mg/ml) [42]. Ginger (Zingiber officinale Rosc.), which has been traditionally used to treat numerous ailments exhibited a significant decrease in the biofilms of *P. aeruginosa.* Zingerone, a bioactive component found in dried ginger root, inhibited biofilm formation and enhanced sensitivity to ciprofloxacin of *P. aeruginosa* PAO1 [43]. Research on bioactives present in citrus plants, showed the inhibition of biofilm formation in *E. coli* O157:H7. Amongst the six bioactives examined, β-sitosterol glucoside suppressed EHEC biofilm substantially. Furthermore, qRT-PCR analysis revealed that the drug suppressed the rssAB gene and mediated regulation of the flagellar master operon flhDC, which inhibited motility and biofilm formation in E. coli O157:H7 [44]. The bioactives in *Rosmarinus officinalis L., Thymus vulgaris L.*, and *Satureja montana* L. essential oils are monoterpene hydrocarbons and phenolic monoterpenes. The research investigation showed that these bioactives impart antibiofilm and anti-adhesive properties to EOs against *Salmonella Typhimurium.* The 2, 3-bis [2-methyloxy-4-nitro-5-sulfophenyl]-2 H-tetrazolium-5-carboxanilide reduction assay was used to assess biofilm inhibition [45]. [46] observed that *S. montana* and *T. vulgaris* essential oils exhibit anti-biofilm activity that is greater than *R. officinalis* EO. Abscissic acid is a plant hormone that acts as QQ molecule against Gram-positive and Gram-negative bacteria. Abscissic acid, its salts, and esters derivatives were tested against *P. acnes, S. aureus, P. aeruginosa, M. furfur*, and other bacterial and fungal species, either alone or in combination with plant extracts (*Morinda citrifolia, Olea europaea var. sylvestris, Curcuma longa*). Abscissic acid was reported to reduce the production of autoinducers by suppressing the expression of Lux-S gene in *P. acnes* cultures. It also reduced expression of inflammation markers (IL-1α) in keratinocytes-bacteria cocultures. In terms of biofilm reduction, they found that abscissic acid (30-60 μg/mL) and plant extracts (2-15 percent by weight) had a synergistic impact against S. aureus and P. acnes growing on abiotic surfaces (glass coupons). All of the combinations tested demonstrated biofilm inhibitory and bacteriostatic properties. Synthetic antibiofilm compounds against *P. aeruginosa* have been designed using natural compounds from sea sponges as scaffolds. The following are some examples of bioactive scaffolds. Bromoageliferinare is a marine sponge-derived alkaloid (Agelas conifer, Agelaceae). Bromoageliferin, a bioactive member to the oroid class, has biological action. TAGE (trans-bromoageliferin analogue 1) and CAGE (cisbromoageliferin analogue 2), two of its counterparts, are also capable of reducing biofilm production. Furthermore, these analogues disperse pre-existing *P. aeruginosa* PAO1 biofilms without impeding their development. CAGE 2 was shown to reduce biofilm mass at concentrations of 400 and 500 μM. The antibiofilm activity of these compounds has to be backed up by a thorough investigation of 2-AI's mechanism of action [47]. Bioactives from probiotic organisms have also gained attention and are used as antimicrobial agents. *Lactobacillus plantarum* 12 exopolysaccharides (L-EPSs) could prominently decrease the synthesis of biofilm matrix of *Shigella flexneri* CMCC51574 and thus can be used as a preventive measure. Furthermore, a cell line study revealed that the compound acts on the adhesion of *S. flexneri* and also hinder its entry into the HT-29 cells. Monosaccharides such as glucose, mannose, xylose, glucuronic acid, galactose, and galactosamine were the components of L-EPS [48]. The bioactive lectins found in the probiotic *Lactobacillus rhamnosus* are utilized to prevent *Salmonella* spp. from forming biofilms. The existence of genes that code for lectin-like proteins was determined by genome sequencing. Lectin like protein (Llp) 1 and 2 are two polypeptides having conserved domains at their N and C termini. Furthermore, LIp (10 g/ml) inhibited 60 percent of *S. Typhimurium* ATCC14028 biofilm formation. Llp1 also reduced the biofilm formation of *E. coli* UTI89 at lower doses. The study also stated that the antibiofilm efficacy of purified lectins was higher than their crude extracts [49].

3.6.2 Synthetic compounds as Quorum Quenchers

Synthetic compounds that can inhibit QS in pathogenic bacteria have been developed. Compounds like Thiazolidinedione (TZD) and its derivatives exhibit QSI activity against *P. aeruginosa*, *V. harveyi* and *Candida albicans*. TZD-C8 ((z)-5-octylidenethiazolidine-2, 4- dione) used at concentration of 20 μM resulted in reduction of biofilm formation in *P. aeruginosa* PAO1 by 70percent. TZD-C8 lead to inhibition of QS signal production in vitro, downregulated the expression of pqsABCDE operon and lasI [50]. Nanoparticles are becoming important due to their antipathogenic activity. Mycofabricated NPs (mfNPs) were formed by incubating silver nanoparticles (AgNPs) with fungi Rhizopus arrhizus BRS-07 biomass. These mfNPs affected lasI and lasR transcription, resulting in a 69 percent to 76 percent reduction in AHL (C4HSL and C12HSL) production. This quorum sensing inhibitor resulted in inhibition of violacein produced by *C. violaceum* CV026 while suppressing the expression of virulence genes by 15–96 percent [51]. A similar research using essential oil-AgNps in treatment of *C. ablicans* infection was reported [52]. Degradable polymer nanofiber loaded with macrocyclic peptide acted as synthetic QSI in the inhibition of *S. aureus.* This QSI was able to inhibit the agr-based QS in S.aureus for about two weeks [53]. Various other synthetic compounds that could exhibit QSI activity are: 1) organosulphur compounds and aryl-l-cysteine sulphoxides inhibited biofilm formation in *Vibrio harveyi* BB152 as well as *V. harveyi* BB170 [54]; 2) metal oxide nanoparticle inhibited *E. coli* and methicillin resistant *S. aureus* (MRSA)[55].

3.6.3 Clinical trial employing QSIs as therapeutics

Pilot scale experiments on furanones have revealed its QS inhibitory activity against various pathogens such as *S. aureus, S. marcescens* , and *P. aeruginosa* which tend to form biofilm on the contact lenses. A reduction of 67-92 percent was observed in bacterial adherence to furanone coated contact lenes when tested for the span of 30 days on animals and 24 hours on humans. The compound was regarded as safe in clinical research experiment [56]. Combination therapy of cinnamaldehyde and tobramycin were effective in reducing load of *Burkholderia cenocepacia* in lungs of infected mice as compared to use of antibiotic alone [57]. These examinations are among the couple of clinical preliminaries utilizing QSIs announced in the public space. A garlic-based oral formulation with QS-inhibitory and calming properties has been licensed and is being used by dental specialists. An inhaler system containing freeze dried AHL-acylase (PvdQ) is used in quenching cystic fibrosis caused by *P. aeruginosa*[58]. This QSI is reported to be stable and can be directly administered into the patients.

3.6.4 Advantages and Disadvantages

Currently there are a few evidences of resistance to QSIs reported. One of the potential mechanisms that counters the QS disruption is via formation of mutant. Cases of *P. aeruginosa* mutant exhibiting higher efflux of brominated furanone (C30) were reported. Under such conditions the efficacy of QSIs is affected, also this leads to emergence of resistances towards the QSIs. The application of enzyme based QSIs that targets the QS signaling molecules outside the bacterial cell are reported as best potential QQ, as they cannot lead to development of resistances. But stability and the efficacy of these enzymes under a diverse set of conditions remains an issue [59]. Another major drawback of using QQ is the lack of sufficient study related to its effect on human microbiota. The QQs can cause dysregulation of multiple functions as well as disturb the QS of microbiota (Firmicutes/Bacteroidetes ratio) thus changing the composition of emerging microbial communities. Organisms like by *Enterobacter, Escherichia, Salmonella* and *Klebsiella* do not possess acyl-HSL synthase, and their QSS operates on AHLs produced by others such as *Y. enterocolitica* [60]. In this situation, QSIs can be anticipated to impact human well being, including metabolic issues, by influencing the stomach microbiome [61]. An elucidatory table (Table 3.2) gives brief description regarding the alternative strategies used to combat AMR.

3.7 Future Perspectives

Traditional antibiotics and antimicrobials have acquired resistance in pathogenic biofilms, posing a difficult issue for clinicians and scientists. Treatment of such persistent biofilms has become time-consuming. Novel antimicrobial approaches are urgently needed to combat the threat of antibiotic resistance. The use of quorum sensing inhibitors or phage therapy in the treatment of antimi-

TABLE 3.2
Basics of alternative strategies to antibiotics

Technology	Actives used	Target Bacteria	Condition	References
Phage Therapy		*P. aeruginosa*	Chronic otitis	(Wright et al., 2009)
		Shigella dysenteriae	Dysentery	(Aishat et al., 2021)
		A.baumannii, P. aeruginosa and *Staphylococcus aureus*	Sepsis	(Soothill, 1992)
		Vibrio parahaemolyticus	Sepsis	(Jun et al., 2014)
		Salmonella typhi	Typhoid	(Kutateladze & Adamia, 2008)
Phage derived lysins	ABgp46	*MDR Acinetobacter baumannii, P. aeruginosa,* and *Salmonella typhimurium*	Inhibition of bacterial density	(Oliveira et al., 2016)
	PlyF307	MDR *A. baumannii*	Bacteremia	(Lood et al., 2015)
Bioengineered chimeric lysins	Artilysins	*P. aeruginosa*	Decolonized P. aeruginosa from gut	(Briers et al., 2014)
	CHAPK	*MRSA*	Eliminated MRSA and dispersed biofilms	(Lood et al., 2015)
Quorum Quencers	Natural:Plant derived	*P. aeruginosa*	Biofilm Inhibition	(Kordbacheh et al., 2017)
	Pistacia atlantica (Anacardiaceae)			
	Ginger: Zingerone	*P. aeruginosa PAO1*	Enhanced sensitivity to ciprofloxacin	(Kumar et al., 2013)
	Citrus plant: β-sitosterol glucoside	*E.coli O157:H7*	Inhibition of biofilm	(Vikram et al., 2013)
	Marine sponge: Agelas conifer	*P. aeruginosa PAO1*	Inhibition of biofilm	(Huigens et al., 2007)
	Bromoageliferinare			
	Bacteria: Lactobacillus rhamnosus	*S.Typhimurium ATCC14028*	Inhibition of biofilm	(Mariya I. Petrova et al., 2016)
	Lectin like protein (Llp) 1 and 2			
	Synthetic: Thiazolidinedione (TZD)	*P. aeruginosa, V. harveyi* and *Candida albicans*	Inhibition of biofilm	(Li et al., 2020)
	Mycofabricated NPs	*C. violaceum* CV026	Inhibition of violacein	(Singh et al., 2015)
	Metaloxide nanoparticle	*E. coli and* methicillin resistant *S. aureus (MRSA)*	Inhibition of biofilm	(Shkodenko and Koshel 2020).
	Nanofiber loaded with macrocyclic peptide	*S. aureus*	Inhibition of QS	(Kratochvil et al., 2017)

crobial resistant organisms are gaining attention. Bioactives are plentiful in natural resources, allowing researchers to investigate a wide library for useful agents. Even though we've made progress in the last few decades in terms of screening antibiofilm bioactives, further research into the bioactive's mechanisms of action is still needed. Furthermore, phase I–IV clinical trials on natural anti-biofilm bioactives necessitate close attention in order to get speedy results. The research on the use of phages and phage-derived proteins in the treatment of bacterial infections, particularly those caused by multidrug-resistant bacteria, is increasingly promising, indicating that phage therapy might be used as an alternative or supplement to antibiotics. The use of bacteriophages for the treatment of bacterial diseases in a post-antibiotic era holds much promise. In order to fully evaluate the potential use of bacteriophages as a treatment option, it is essential that all aspects of the biology of the phages be fully understood. However, new studies on immunomodulatory effects, host range, and the possibility for horizontal gene transfer show that a greater knowledge of the interplay between phage, microbiota, and human host is needed before phage treatment can be widely used. Biotechnological advancements have broadened the range of possible phage therapies to include bioengineered phages and purified phage lytic proteins, among other approaches. Thus, the therapeutic use of natural anti-biofilm agents and bacteriophages has huge potential in the management of drug-resistant bacterial infections and will support health care industries all over the world. The chapter discusses techniques to eradicating pathogenic biofilm development and associated illnesses by combining bioactives and phages with antibiotics as a preventative and therapeutic tool.

Bibliography

[1] Sabrina Mühlen and Petra Dersch. Anti-virulence strategies to target bacterial infections. *How to overcome the antibiotic crisis*, pages 147–183, 2015.

[2] Paramita Sarkar, Venkateswarlu Yarlagadda, Chandradhish Ghosh, and Jayanta Haldar. A review on cell wall synthesis inhibitors with an emphasis on glycopeptide antibiotics. *Medchemcomm*, 8(3):516–533, 2017.

[3] Christina R Bourne. Utility of the biosynthetic folate pathway for targets in antimicrobial discovery. *Antibiotics*, 3(1):1–28, 2014.

[4] David C Hooper. Quinolone mode of action. *Drugs*, 49(2):10–15, 1995.

[5] Lisa S McCoy, Yun Xie, and Yitzhak Tor. Antibiotics that target protein synthesis. *Wiley Interdisciplinary Reviews: RNA*, 2(2):209–232, 2011.

[6] Tom Defoirdt, Gilles Brackman, and Tom Coenye. Quorum sensing inhibitors: how strong is the evidence? *Trends in microbiology*, 21(12):619–624, 2013.

[7] Vipin Chandra Kalia. Quorum sensing inhibitors: an overview. *Biotechnology advances*, 31(2):224–245, 2013.

[8] Marianne Frieri, Krishan Kumar, and Anthony Boutin. Antibiotic resistance. *Journal of infection and public health*, 10(4):369–378, 2017.

[9] Rogerio Caldeira Rodrigues Malta, Gustavo Luis de Paiva Anciens Ramos, and Janaína dos Santos Nascimento. From food to hospital: we need to talk about acinetobacter spp. *Germs*, 10(3):210, 2020.

[10] Stephanie Fletcher. Understanding the contribution of environmental factors in the spread of antimicrobial resistance. *Environmental health and preventive medicine*, 20(4):243–252, 2015.

[11] Arthur Neuberger, Dijun Du, and Ben F Luisi. Structure and mechanism of bacterial tripartite efflux pumps. *Research in microbiology*, 169(7-8):401–413, 2018.

[12] áM Baroud, I Dandache, GF Araj, R Wakim, S Kanj, Z Kanafani, M Khairallah, A Sabra, M Shehab, G Dbaibo, et al. Underlying mechanisms of carbapenem resistance in extended-spectrum β-lactamase-producing klebsiella pneumoniae and escherichia coli isolates at a tertiary care centre in lebanon: role of oxa-48 and ndm-1 carbapenemases. *International journal of antimicrobial agents*, 41(1):75–79, 2013.

[13] Pradeep Reddy Marri, Weilong Hao, and G Brian Golding. The role of laterally transferred genes in adaptive evolution. *BMC evolutionary biology*, 7(1):1–14, 2007.

[14] Patricia L Foster. Stress-induced mutagenesis in bacteria. *Critical reviews in biochemistry and molecular biology*, 42(5):373–397, 2007.

[15] Gerard D Wright. Antibiotic resistance in the environment: a link to the clinic? *Current opinion in microbiology*, 13(5):589–594, 2010.

[16] Peter A Lambert. Bacterial resistance to antibiotics: modified target sites. *Advanced drug delivery reviews*, 57(10):1471–1485, 2005.

[17] Jan AJW Kluytmans, Ilse TMA Overdevest, Ina Willemsen, Marjolein FQ Kluytmans-Van Den Bergh, Kim Van Der Zwaluw, Max Heck, Martine Rijnsburger, Christina MJE Vandenbroucke-Grauls, Paul HM Savelkoul, Brian D Johnston, et al. Extended-spectrum β-lactamase–producing escherichia coli from retail chicken meat and humans: comparison of strains, plasmids, resistance genes, and virulence factors. *Clinical Infectious Diseases*, 56(4):478–487, 2013.

[18] V Besnard, M Federighi, and JM Cappelier. Development of a direct viable count procedure for the investigation of vbnc state in listeria monocytogenes. *Letters in applied Microbiology*, 31(1):77–81, 2000.

[19] PM Bennett. Plasmid encoded antibiotic resistance: acquisition and transfer of antibiotic resistance genes in bacteria. *British journal of pharmacology*, 153(S1):S347–S357, 2008.

[20] Rajesh R Uchil, Gurdeep Singh Kohli, Vijay M KateKhaye, and Onkar C Swami. Strategies to combat antimicrobial resistance. *Journal of clinical and diagnostic research: JCDR*, 8(7):ME01, 2014.

[21] CDC. CDC launches networks to boost global response to antibiotic resistance — CIDRAP, 2021.

[22] World Health Organization. Antimicrobial resistance, 2021.

[23] Kim Lewis. Platforms for antibiotic discovery. *Nature reviews Drug discovery*, 12(5):371–387, 2013.

[24] Stefano Donadio, Sonia Maffioli, Paolo Monciardini, Margherita Sosio, and Daniela Jabes. Antibiotic discovery in the twenty-first century: current trends and future perspectives. *The Journal of antibiotics*, 63(8):423–430, 2010.

[25] Environment Federal Ministries of Agriculture and Health. Nigeria AMR Action Plan. 2017.

[26] Moi Lin Ling, Anucha Apisarnthanarak, and Gilbert Madriaga. The burden of healthcare-associated infections in southeast asia: a systematic literature review and meta-analysis. *Clinical Infectious Diseases*, 60(11):1690–1699, 2015.

[27] Niels Høiby, Oana Ciofu, Helle Krogh Johansen, Zhi-jun Song, Claus Moser, Peter Østrup Jensen, Søren Molin, Michael Givskov, Tim Tolker-Nielsen, and Thomas Bjarnsholt. The clinical impact of bacterial biofilms. *International journal of oral science*, 3(2):55–65, 2011.

[28] Ido Yosef, Ruth Kiro, Shahar Molshanski-Mor, Rotem Edgar, and Udi Qimron. Different approaches for using bacteriophages against antibiotic-resistant bacteria. *Bacteriophage*, 4(1):19549–54, 2014.

[29] George PC Salmond and Peter C Fineran. A century of the phage: past, present and future. *Nature Reviews Microbiology*, 13(12):777–786, 2015.

[30] Patrick Jault, Thomas Leclerc, Serge Jennes, Jean Paul Pirnay, Yok-Ai Que, Gregory Resch, Anne Françoise Rousseau, François Ravat, Hervé Carsin, Ronan Le Floch, et al. Efficacy and tolerability of a cocktail of bacteriophages to treat burn wounds infected by pseudomonas aeruginosa (phagoburn): a randomised, controlled, double-blind phase 1/2 trial. *The Lancet Infectious Diseases*, 19(1):35–45, 2019.

[31] Ryohei Watanabe, Tetsuya Matsumoto, Go Sano, Yoshikazu Ishii, Kazuhiro Tateda, Yoshinobu Sumiyama, Jumpei Uchiyama, Shingo Sakurai, Shigenobu Matsuzaki, Shosuke Imai, et al. Efficacy of bacteriophage therapy against gut-derived sepsis caused by pseudomonas aeruginosa in mice. *Antimicrobial agents and chemotherapy*, 51(2):446–452, 2007.

[32] Maciej Żaczek, Beata Weber-Dabrowska, Ryszard Miedzybrodzki, Marzanna Łusiak-Szelachowska, and Andrzej Górski. Phage therapy in poland–a centennial journey to the first ethically approved treatment facility in europe. *Frontiers in Microbiology*, 11:1056, 2020.

[33] Stephen T Abedon, Sarah J Kuhl, Bob G Blasdel, and Elizabeth Martin Kutter. Phage treatment of human infections. *Bacteriophage*, 1(2):66–85, 2011.

[34] Oluwafolajimi Adesanya, Tolulope Oduselu, Oluwawapelumi Akin-Ajani, Olubusuyi M Adewumi, and Olusegun G Ademowo. An exegesis of bacteriophage therapy: An emerging player in the fight against anti-microbial resistance. *AIMS microbiology*, 6(3):204, 2020.

[35] Zhong Peng, Dazhi Jin, Hyeun Bum Kim, Charles W Stratton, Bin Wu, Yi-Wei Tang, and Xingmin Sun. Update on antimicrobial resistance in clostridium difficile: resistance mechanisms and antimicrobial susceptibility testing. *Journal of clinical microbiology*, 55(7):1998–2008, 2017.

[36] George Tetz and Victor Tetz. Bacteriophage infections of microbiota can lead to leaky gut in an experimental rodent model. *Gut pathogens*, 8(1):1–4, 2016.

[37] Jan Borysowski and Andrzej Górski. Is phage therapy acceptable in the immunocompromised host? *International Journal of Infectious Diseases*, 12(5):466–471, 2008.

[38] Eric V Granowitz and Richard B Brown. Antibiotic adverse reactions and drug interactions. *Critical care clinics*, 24(2):421–442, 2008.

[39] Jianlong Yan, Jiaoxiao Mao, and Jianping Xie. Bacteriophage polysaccharide depolymerases and biomedical applications. *BioDrugs*, 28(3):265–274, 2014.

[40] Nicola Principi, Ettore Silvestri, and Susanna Esposito. Advantages and limitations of bacteriophages for the treatment of bacterial infections. *Frontiers in pharmacology*, 10:513, 2019.

[41] Qian Jiang, Jiashun Chen, Chengbo Yang, Yulong Yin, and Kang Yao. Quorum sensing: a prospective therapeutic target for bacterial diseases. *BioMed Research International*, 2019, 2019.

[42] H Kordbacheh, F Eftekhar, and SN Ebrahimi. Anti-quorum sensing activity of pistacia atlantica against pseudomonas aeruginosa pao1 and identification of its bioactive compounds. *Microbial Pathogenesis*, 110:390–398, 2017.

[43] Lokender Kumar, Sanjay Chhibber, and Kusum Harjai. Zingerone inhibit biofilm formation and improve antibiofilm efficacy of ciprofloxacin against pseudomonas aeruginosa pao1. *Fitoterapia*, 90:73–78, 2013.

[44] Amit Vikram, GK Jayaprakasha, Ram M Uckoo, and Bhimanagouda S Patil. Inhibition of escherichia coli o157: H7 motility and biofilm by β-sitosterol glucoside. *Biochimica et Biophysica Acta (BBA)-General Subjects*, 1830(11):5219–5228, 2013.

[45] Anu Iswarya Jaisankar, AS Smiline Girija, Shoba Gunasekaran, and J Vijayashree Priyadharsini. Molecular characterisation of csga gene among esbl strains of a. baumannii and targeting with essential oil compounds from azadirachta indica. *Journal of King Saud University-Science*, 32(8):3380–3387, 2020.

[46] Hanene Miladi, Donia Mili, Rihab Ben Slama, Sami Zouari, Emna Ammar, and Amina Bakhrouf. Antibiofilm formation and anti-adhesive property of three mediterranean essential oils against a foodborne pathogen salmonella strain. *Microbial Pathogenesis*, 93:22–31, 2016.

[47] Robert W Huigens, Justin J Richards, Gina Parise, T Eric Ballard, Wei Zeng, Rajendar Deora, and Christian Melander. Inhibition of pseudomonas aeruginosa biofilm formation with bromoageliferin analogues. *Journal of the American Chemical Society*, 129(22):6966–6967, 2007.

[48] Peter M Hawkey and Annie M Jones. The changing epidemiology of resistance. *Journal of antimicrobial chemotherapy*, 64(suppl_1):i3–i10, 2009.

[49] Zalak Rameshbhai Matariya, Kaushik Kishorbhai Lodhiya, and Rajkumar G Mahajan. Environmental correlates of undernutrition among children of 3–6 years of age, rajkot, gujarat, india. *Journal of family medicine and primary care*, 5(4):834, 2016.

[50] Wen-Ru Li, Tao-Hua Zeng, Xiao-Bao Xie, Qing-Shan Shi, and Cai-Ling Li. Inhibition of the pqsabcde and pqsh in the pqs quorum sensing system and related virulence factors of the pseudomonas aeruginosa pao1 strain by farnesol. *International Biodeterioration & Biodegradation*, 151:104956, 2020.

[51] Braj R Singh, Brahma N Singh, Akanksha Singh, Wasi Khan, Alim H Naqvi, and Harikesh B Singh. Mycofabricated biosilver nanoparticles interrupt pseudomonas aeruginosa quorum sensing systems. *Scientific reports*, 5(1):1–14, 2015.

[52] RA Hikmet and NN Hussein. Mycosynthesis of silver nanoparticles by candida albicans yeast and its biological applications. *Archives of Razi Institute*, 76(4):857–869, 2021.

[53] Michael J Kratochvil, Tian Yang, Helen E Blackwell, and David M Lynn. Nonwoven polymer nanofiber coatings that inhibit quorum sensing in staphylococcus aureus: toward new nonbactericidal approaches to infection control. *ACS infectious diseases*, 3(4):271–280, 2017.

[54] SH Kasper, D Samarian, AP Jadhav, AH Rickard, RA Musah, and NC Cady. S-aryl-l-cysteine sulphoxides and related organosulphur compounds alter oral biofilm development and ai-2-based cell–cell communication. *Journal of applied microbiology*, 117(5):1472–1486, 2014.

[55] Liubov Shkodenko, Ilia Kassirov, and Elena Koshel. Metal oxide nanoparticles against bacterial biofilms: perspectives and limitations. *Microorganisms*, 8(10):1545, 2020.

[56] Zohra Khatoon, Christopher D McTiernan, Erik J Suuronen, Thien-Fah Mah, and Emilio I Alarcon. Bacterial biofilm formation on implantable devices and approaches to its treatment and prevention. *Heliyon*, 4(12):e01067, 2018.

[57] Omar M El-Halfawy, Marwa M Naguib, and Miguel A Valvano. Novel antibiotic combinations proposed for treatment of burkholderia cepacia complex infections. *Antimicrobial Resistance & Infection Control*, 6(1):1–5, 2017.

[58] Franziska S Birmes, Ruth Säring, Miriam C Hauke, Niklas H Ritzmann, Steffen L Drees, Jens Daniel, Janina Treffon, Eva Liebau, Barbara C Kahl, and Susanne Fetzner. Interference with pseudomonas aeruginosa quorum sensing and virulence by the mycobacterial pseudomonas quinolone signal dioxygenase aqdc in combination with the n-acylhomoserine lactone

lactonase qsda. *Infection and Immunity*, 87(10):e00278–19, 2019.

[59] Rodolfo García-Contreras, Toshinari Maeda, and Thomas K Wood. Can resistance against quorum-sensing interference be selected? *The ISME journal*, 10(1):4–10, 2016.

[60] YN Nguyen, Haiqing Sheng, Rambabu Dakarapu, John R Falck, Carolyn J Hovde, and Vanessa Sperandio. The ahl-synthase yeni from yersinia enterocolitica modulates virulence gene expression in enterohemorrhagic e. coli (ehec). *Infection and Immunity*, 2013.

[61] Hannah McCarthy, Justine K Rudkin, Nikki S Black, Laura Gallagher, Eoghan O'Neill, and James P O'Gara. Methicillin resistance and the biofilm phenotype in staphylococcus aureus. *Frontiers in cellular and infection microbiology*, 5:1, 2015.

4

Antimicrobials in Growth and Development

Ujjayni Saha

Symbiosis School of Biological Sciences, Symbiosis International (Deemed University) Pune, 412115, India

Ranjika Bhattacharya

Symbiosis School of Biological Sciences, Symbiosis International (Deemed University) Pune, 412115, India

Sunil D. Saroj

Symbiosis School of Biological Sciences, Symbiosis International (Deemed University) Pune, 412115, India

CONTENTS

4.1 Introduction

Antimicrobials are a diverse group of compounds frequently used to treat infections caused by pathogenic microorganisms. In 1928, Alexander Fleming made the breakthrough discovery of penicillin, the first known antibiotic, for which he was awarded the Nobel Prize in Physiology/Medicine in 1945. Over the years, this was followed by the unearthing of many other findings that have immensely helped to combat a broad range of infectious diseases [1]. Antimicrobials include a range of antibiotics, antiparasitic, antifungals, and antiviral drugs, which help eradicate many pathogens. However, due to indiscriminate and irrational use of these antimicrobials, these pathogens change over time and no longer respond to these drugs, failing to clear infections, giving rise to antimicrobial resistance (AMR), in which microorganisms mutate or undergo certain adaptations to render treatments futile. This increases the risk of disease spread and death in humans, animals, and plants. The growing threat of antimicrobial resistance to health care, as well as the resulting lack of access to effective antimicrobials, is a global problem [2]. There are over 670,000 infections due to antibiotic-resistant bacteria in Europe each year, nearly with a death toll of 33,000. This indicates that the dissemination of resistant clones in healthcare systems has led to fewer treatment options (https://apps.who.int/iris/handle/10665/351141). More than 2.8 million antibiotic-resistant illnesses occur in the United States each year. According to the CDC's 2019 Antibiotic Resistance (AR) Threats Report, more than 35,000 people die as a result. Antibiotic resistance knows no bounds, endangering modern medicine and our capacity to cure common diseases. The COVID-19 pandemic experience has demonstrated that additional effort is required to establish robust and adaptable health systems capable of recognising, containing, and halting a pandemic wherever it begins (https://www.cdc.gov/drugresistance/about.html). Regulatory policies are deficient in developing nations and middle/low-income countries such as India and Vietnam, where it is relatively easy to access over-the-counter (OTC) antibiotics [3]. This further encourages people to perform self-treatment, which can have very adverse effects in the long run [4]. These statistics indicate that the dissemination of resistant clones or "superbugs" in healthcare systems has led to very few options for treatment. The fast emergence of "superbugs" has resulted in their ability to be resistant to various medications, rendering them ineffectual. One of the many examples is methicillin resistance in *Staphylococcus aureus* (MRSA) which has a high mortality rate worldwide. Conventionally, glycopeptides were used to treat this infection; however, over time, MRSA has become less susceptible to vancomycin owing to various cellular changes such as thicker cell wall [5] and higher amount of glutamine non-aminated muropeptides that cause lesser peptidoglycan cross-linking [6]. Other examples include HIV drug resistance (HIVDR) which occurs when HIV mutations evolve as a result of viral replication in people using antiretroviral medications, carbapenem-resistant *Enterobacteriaceae*, which confer broad resistance to β-lactam antibiotics including "last-line" carbapenems [7], multidrug-resistant *Pseudomonas aeruginosa* which show resistance to 3 or more classes of antipseudomonal agents [8], etc. Antimicrobial misuse in many industries is undoubtedly one of the most essential and fundamental causes of AMR; nevertheless, there are other multidisciplinary elements at play that significantly contribute to this issue. Apart from being used by humans, these drugs are also very commonly used for growth promotion, nutritional improvement, prophylaxis (prevention of infection), and treatment in groups of food-producing animals since the 1990s. For example, penicillin was widely used to treat a range of Gram-positive infections in wounded soldiers, such as septicaemia. At the same time, it was also used to treat bovine mastitis by veterinarians [9]. Antimicrobial resistance in zoonotic enteropathogens (e.g., *Salmonella, Campylobacter, Yersinia,* and some strains of *Escherichia coli,* such as serotype O157:H7) and commensals (e.g., *Enterococci, E. coli*) is of particular concern to human health because these bacteria are most likely to be transmitted to humans through the food chain [10]. Additionally, residues of antimicrobials can be found in animal waste that contaminates the surrounding environment, such as soil, plants, water sources, etc. The additive effect of the above conditions results in increased selection pressure on pathogens, i.e., the microbes that have acquired resistance genes against the antimicrobials will have a relatively more significant chance of survival. In terms of human health, according to [11], there are two primary categories of harmful impacts from the presence of antibiotic residues in food of animal origin. Hypersensitive (immunological, allergic) responses are the first. Immunological responses might range from a minor rash to life-threatening situations like anaphylactic shock. The development of antibiotic resistance is the second sort of detrimental consequence. To combat this, many countries have banned or imposed stringent guidelines on the use of antimicrobials in food animals [12]. In this chapter, we will mainly be looking at the concept of antimicrobial resistance, antimicrobials used in food animals and humans for disease prevention, growth promotion and treatment, and how this plays a pivotal role in the increase in AMR worldwide.

4.2 Antimicrobial resistance

AMR is the process by which a microbe evolves to become increasingly or entirely resistant to an antimicrobial treatment that could previously effectively treat it. Antibiotic resistance, which primarily pertains to bacteria and antibiotics in a clinical setting, refers to a microorganism's resistance to an antibiotic caused by antibiotic usage. AMR is on the rise, particularly in the vicinity of healthcare-associated infections (HAIs), also known as nosocomial infections [13]. As AMR is on the rise, the effectiveness of antimicrobials has declined extensively, making it difficult for physicians. In this case, they are left with no choice but to administer last-resort antibiotics like polymyxins and carbapenems, which are not readily available in many developing countries and have specific side effects, including nephrotoxicity [14]. This stands to be a big problem for patients with a compromised immune system, such as those undergoing chemotherapy, surgery, dialysis, etc., or those with chronic conditions like asthma, diabetes, and rheumatoid arthritis [4]. Apart from the reckless use of antibiotics, there are other reasons for the emergence of AMR involving unnecessary prescriptions by physicians for extended durations [15], improper sanitation measures, poor control of infection, and inappropriate food-handling [16]. AMR has reached its peak point, and now researchers claim that we are in the post-antibiotic era. Recent reports have focused on this tendency which includes the emergence of various plasmid-mediated colistin resistance genes in humans and animal pathogens [11]. Antimicrobial resistance in food animals has a remarkable impact on animal health and can directly contribute to its spread in humans. A study found that low-middle income countries account for up to a third of the global increase (67 percent) in antibiotic consumption in food animals between 2010 and 2030, highlighting the burden of infectious diseases in developing countries as well as the intensive use of antimicrobials to combat AMR [17].

4.2.1 Drivers of AMR

The finding that microbial antibiotic resistance genes are transmissible and opportunistic, spreading from organism to organism, is pivotal in the knowledge of microorganisms and their role in infection. Bacterial resistance can occur for various causes, including a mutation of the antibiotic-target gene in the bacterial chromosome or elsewhere due to the insertion of extra-chromosomal DNA. The transfer of plasmids and conjugative transposons during cell-to-cell contact during conjugation is the most effective and has the most significant impact on the spread of antibiotic resistance [18]. Both Gram-negative and Gram-positive bacteria may now clearly fulfil the evolutionary challenge of fighting antimicrobial treatment by obtaining pre-existing resistance determinants from the bacterial gene pool. This is accomplished by the coordinated actions of mobile genetic elements like insertion sequences, gene cassettes or integrons, and transposons which can move inside or across DNA molecules [19].

4.2.1.1 AMR mechanisms can fall into four different groups

- Modification of drug targets- The bacterial cells contain many components that can act against antimicrobial agents. For example, a change in the structure of PBP's (penicillin-binding proteins) might reduce the drug binding activity or, moreover, totally inhibit drug binding [20].

- Drug uptake limitation- Naturally, the ability to limit antimicrobial agents' uptake is different across bacteria. For example, in the *Enterobacteriaceae* family, members become resistant by stopping porin production or reducing the number of certain porins. Reducing the porin number works as a mechanism for resistance to carbapenems [21].

- Active drug efflux- Encoded genes in chromosomes for efflux pumps are present in bacteria. Successively fewer are expressed while others are overexpressed or induced. The efflux pump works by expelling toxic substances from the bacterial cell. Many efflux pump families are single-component pumps that transport substrates across the cytoplasmic membrane [22].

- Drug inactivation- There are two ways in which this method works. The first is by degradation of the drug, and the second is by the transfer of the chemical group to the drug [23]. Drug inactivation directed by transferring a chemical group to the drug is often done with phosphoryl, acetyl, and adenyl groups [22].

4.2.1.2 What is making AMR grow so extensively?

Biological factors- Antibiotic resistance can happen spontaneously through bacterial evolution and mutations [24]. Bacteria can acquire resistance-inducing genetic material through various mechanisms, including transformation, transposition, and conjugation, which is referred to as horizontal gene transfer (HGT). Plasmid-mediated transmission is the most prevalent method, while bacteriophage-mediated transmission is rare [25].

- Overuse/Misuse of antimicrobials- This is a common factor leading to AMR widely. Various surveys across the world have stated that humans take antibiotics for the common cold, which can lead to resistance to the particular drug taken [26]. Moreover, developing countries do not have proper facilities for correctly diagnosing diseases and infections, and patient management is also improper, including the prescription of antibiotics. These antibiotics have quality issues; still they are sold at over-the-counter stores [27].

- Use of antimicrobials in agriculture- The use of antimicrobials in agriculture is one of the key promotors for spreading AMR [28]. Antibiotics are used to treat diseased animals and have also been used extensively for prophylaxis of healthy animals that grow in subtherapeutic levels and increase feed efficacy [29]. For example, pigs are fed with colistin to promote their growth, which is the last resort antibiotic for treatment in humans; and the resistance of this drug can spread from pigs to humans via different routes [30]. According to many studies, AMR bacteria and their AMR determinants can be transmitted from food animals to people through direct contact and/or animal products [3, 31].

- Lack of substitute antimicrobials- In contrast to affluent countries, which have effectively prohibited the use of antimicrobials as growth promoters and replaced them with other growth promoters and good practices without affecting the performance of their livestock sectors, developing countries lack alternatives. This would be difficult to achieve in underdeveloped countries because people are more likely to buy antimicrobials on the black market, where they may be of low quality, worsening the problem and leading to a significant rise in illness, with associated morbidity and mortality [32].

- Travel routes- Various studies have interpreted that the travel routes of humans, animals, migratory birds, and goods have facilitated the spread of AMR. Humans travel across the globe, and once they get affected and get back to their country, there is a high probability of them spreading AMR organisms. Humans migrate over the world, and once they get infected and return to their home country, there is a strong likelihood that they will disseminate AMR genes [33].

- The knowledge gap- Many countries are still unaware of AMR risks. Furthermore, there is a lack of proper 'One Health' integration between animal and human healthcare. Only a few countries follow global surveillance systems to collect data about AMR across various sectors [34]. The knowledge gap needs further attention and interdisciplinary efforts from multiple sectors like medicine, veterinary, agriculture, animal feed producing industries, and most importantly, the consumers.

4.2.2 Transmission of AMR from animals to humans or vice versa by different routes

The several steps involved in the prescription, production, distribution, dispensing, and finally, consumption of the drug by humans and its use in animal production frequently contribute to AMR's emergence [35]. Antibiotic-resistant pathogens can contaminate various surroundings. AMR bacteria spread across human populations in the same way as infectious illnesses do, namely by direct contact from person to person, or from animal to person (via scratches or bites), and vertically from mother to unborn child. Meat derived from animals containing these pathogens can affect humans consuming it, whereas humans can also spread AMR to animals while handling livestock and poultry (Figure 4.1). For example, isolation of streptogramin-resistant *E. faecium* was done from Dutch chicken farmers and also from their chicken faeces [36]. Such strains can also reach the environment via sewage and manure, and the food consumed from such environments can lead to the widespread dissemination of AMR. Particulate matter from beef cattle feeds enabled the dissemination of antimicrobial-resistant genes by airborne transmission [12]. One of the most well-known cases under this category is that of LA-MRSA (livestock-associated methicillin-resistant *S. aureus*) CC (clonal complex) 398, which exclusively colonizes in food animals. However, it was seen that it was also able to cause infection in people that were actively involved in handling livestock and those that were living in close vicinity to these infected animals. This can be attributed to a "spillover" event and constant adaptations because the pathogen was able to cross the animal-human species barrier and cause in-

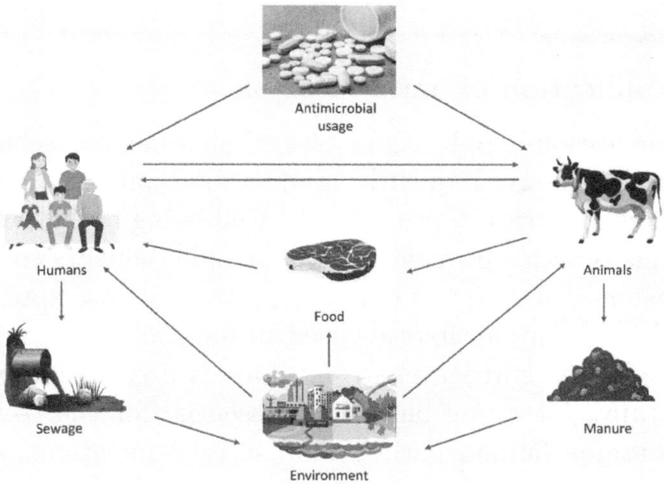

FIGURE 4.1
Transmission routes of AMR: Different transmission routes of AMR amongst humans, animals, and the environment. Humans and animals can directly contract AMR from each other or modes like food, environment, manure, sewage, etc.

fection in humans by evading phagocytosis. Recent studies demonstrate that this pathogen has now been introduced into hospital settings. It is now capable of causing nosocomial infections, which makes it a significant threat to human health [37]. Rodents, wild animals, and birds can serve as a reservoir for AMR Gram-negative bacteria, passing them on to grazing areas, animal feed, or directly passing through their excreta [38]. Vancomycin-resistant enterococci (VRE) have been found in pet animals and rodents, leading to widespread dissemination in humans [39]. In many countries, human faeces are used as fertilisers. They can spread AMR organisms/genes to vegetables. Multiple antibiotic-resistance E.coli strains were found in market vegetables in London [40]. *Salmonella* spps. and *Campylobacter* spps. infection by zoonoses can reach humans via the routes of the food chain through contaminated meat and poultry [40]. Fish farming is integrated with wastewater, sewage, agriculture, and manure in the fishing industry. Their residues get incorporated with the fish feed, spreading to the whole water body [40].

4.3 Application of antimicrobials

There are various applications of antimicrobials as seen in Table 4.1. Antibiotics are frequently used in food-animal production in developing countries to promote the well-being and growth of animals. This practice provides some economic benefits to producers and consumers at large. They also prevent disease spread among animal lots. Many antibiotics used in food animals are currently the same as the surrogates for antibiotics used in human medicinal therapy. These antibiotics are essential for the treatment of many common human pathogenic bacterial infections, as well as anti-infection agents in other treatments such as major surgical operations, chemotherapy for cancer, organ transplantation, etc [17].

4.3.1 Disease prevention

Guidelines for antibiotic prophylaxis in the medical field can only be found for a few human medical conditions like acute rheumatic fever, congenital heart anomalies, asplenia, and sickle cell disease vesicoureteral reflux, pertussis, and meningococcal exposure, latent tuberculosis infection [37]. Also, in the surgical field, national guidelines recommend that the surgical antimicrobial prophylaxis should be initiated within 1 hour before incision at the start of surgery and discontinued within 24 hours postoperatively for most procedures and for cardiac surgery within 48 hours [41]. The physicians' involvement is critical in avoiding and controlling infections that frequently need quick source control and adequate antibiotic therapy, and they are directly accountable for the infection's prognosis. Surgeons must enhance surgical care and avoid improper antibiotic administration in surgery since patients are at risk of getting antibiotics for prophylactic and therapeutic purposes for lengthy periods of time, leading to antibiotic resistance [42]. "Prophylaxis" is also when sub-therapeutic doses of antimicrobials are given in animal feed to seemingly healthy animals that are not exhibiting any symptoms of disease; however, they might be at risk of illness in the future. This is especially significant in the veterinary field since studies show that resistant organisms or genes can be passed from animals to people [43]. Many products used for prophylaxis, such as bambermycins, bacitracin, carbadox, have a minimum or nil application in human medicine. However, these prophylactic drugs and therapy are often closely related to antibiotics used in human medicine. The classes used for prophylaxis include β-lactams like penicillin and cephalosporins; tetracyclines, sulphonamides, macrolides, quinolones (including fluoroquinolones) [44]. Antibiotics like oxytetracycline, enrofloxacin, ampicillin, sparfloxacin, and ciprofloxacin have been used in aquaculture for prevention and therapy, as well as to promote larval survival. In one investigation, 90 *Vibrio* spp. isolates and 7 *Aeromonas* spp. Isolates from ponds and hatcheries on the East coast of India were found to be 100 percent resistant to ampicillin, and almost half of the isolates were resistant to ciprofloxacin, ceftriaxone, kanamycin, and furazolidone [45]. As a result, it is critical to minimise the use of prophylactic antibiotics in aquaculture and focus on proper implementation standards. Antibiotic prophylaxis is not successful in all therapeutic contexts and can harm individual animals (for example, when antimicrobial usage causes superinfection) or the population as a whole (for example, when antimicrobial use leads to dissemination of antimicrobial-resistant bacteria) [46].

4.3.2 Disease treatment

Infection and illness therapy has dramatically improved because of the fast development of antibacterial compounds during the last 100 years. "Therapy" refers to administering antimicrobials in animals that are manifesting symptoms of clinical disease [44]. Another term commonly used is "metaphylaxis," which means giving low doses of antimicrobials to a group of animals for an extended period. For metaphylaxis, around 41 percent of feedlots used antimicrobials such tilmicosin, florfenicol, and oxytetracycline. Sometimes, this term may also refer to the mass-medication of a group of diseased animals [12]. Antimicrobials are also given to cattle on feedlots for a number of reasons, including the prevention or treatment of respiratory illness outbreaks, the management of liver abscesses, etc. In 1910,Ehrlich created Salvarsan, a syphilis treatment, introduced as one of the world's first antimicrobial drugs. Sulphonamides were synthesised by Domagk and other researchers in 1935. Several more antimicrobial drug families were developed during the next two decades, ushering in the "golden age of antimicrobial chemotherapy." One such antibiotic is streptomycin, an aminoglycoside antibiotic discovered in 1944 from the soil bacterium *Streptomyces griseus.* Other soil bacteria created chloramphenicol, rifampicin, tetracyclines, macrolides, and glycopeptides. The first quinolone antibiotic, nalidixic acid, was developed in 1962, followed by cephalosporins, which were discovered in the 1960s and soon gained popularity among doctors. Since then, antimicrobial medicines have improved in terms of intrinsic efficacy and spectrum of activity [47]

- Bacterial infections- Mainly, many serious infections are exclusively caused by bacteria. As they are distinctly related to humans, there are various selective targets to control them. They have a cell wall made of peptidoglycan, which is absent in humans, and thus this makes them a suitable target. Similarly, other components like bacterial ribosomes enzymes required for nucleic acid synthesis are ideal as they vary in humans and animals.

- Bacterial cell wall damage can be caused by beta-lactams like penicillin cephalosporins; protein synthesis prevention by macrolides like erythromycin, chloramphenicol, clindamycin, tetracyclines; prevention of nucleic acid synthesis by fluoroquinolones, rifampicin; inhibiting metabolic pathway by sulphonamides, trimethoprim; membrane structure disruption by colistin, daptomycin [48].

- Fungal infections- As fungi are more closely related to humans compared to bacteria, they have had comparatively less time for differential evolution, so only a few targets are available like cell wall-glucan inhibitors, which include echinocandins (micafungin, caspofungin), cell membrane-ergosterol binders like polyenes (amphotericin, nystatin), cell membrane-ergosterol binders like azoles (itraconazole, ketoconazole, voriconazole, etc.) [49].

- Viral infections- Viruses are simple organisms with no self-metabolism and need a host to survive, making them intracellular pathogens. They can be treated by inhibiting viral DNA polymerase using acyclovir matrix using amantadine and rimantadine. However , neuraminidase inhibitors like oseltamivir zanamivir are currently used widely [13].
- Parasitic infections- There are currently fewer treatments for protozoa and helminths. Usually, preventive measures are taken to reduce the incidence of mosquitoes, like removal of still water, bed nets, full-length clothes, etc. It is advised that travellers going to malaria-endemic areas take prophylaxis [13].

4.3.3 Growth promotion

As per National Committee for Clinical Laboratory Standards (NCCLS), the term "growth promotion" refers to the administration of low, sub-therapeutic concentrations of antimicrobials over an extended period, generally the entire lifetime of the animal, to enhance the overall performance and efficiency of animals such as pigs, chickens, etc. [40]. These phenomena were first observed in the 1940s when poultry that was given feed containing tetracycline was seen to be growing much faster as compared to those that were given feed devoid of antibiotics [50]. From this point onwards, various antimicrobials were found to have growth promotion potential in livestock, including chickens, swine, and cattle. The FDA (Food and drug administration) in the USA approved the use of antimicrobial growth promoters in animals (AGP) in 1951 (https://www.fda.gov/about-fda/histories-product-regulation/animal-health-consumer-protection). It was also used in pigs, resulting in overall weight gain [21]. The inflammatory and pro-inflammatory mediators can affect animals by reducing their growth, milk production and metabolic activities. Thus, the anti-inflammatory antibiotics, especially macrolides, proved beneficial for overall animal health and development [51]. However, on the other hand, antibiotic use in animals for growth promotion has widely led to the emergence of AMR strains. According to WHO, in the past decade, the deaths caused due to

AMR surpassed the number of deaths caused by HIV, influenza and traffic accidents [52]. Avoparcin is used as a glycopeptide antibiotic used as a feed additive. It is associated with the appearance of the avoparcin-resistant strain, which also cross-tolerated vancomycin, a drug used in humans [53]. The United Kingdom first recognised the issue of antimicrobials and growth promoters in the 1960s. After this, the SWANN commission recommended that antibiotic use for animal growth promotion should be banned. Later in 1999 the European Union made regulations in the agriculture sector by prohibiting four classes of different antibiotics used in growth promotion. Then in 2006, it banned courses of antibiotics used for growth promotion in feed producing animals [50] (Table 4.1). Alternatives to antibiotic growth promotors include in-feed enzymes, which help break down components in feed such as proteins, phylates, glucans that are hard to digest. They can be produced by fermenting products from bacteria and fungi. Probiotics can be incorporated to enhance gut health. They strongly influence intestinal metabolism by increasing vitamin B12 and bacteriocin production. Moreover, competitive exclusion products can be used in animal feeds which colonise the animal gut and prevent infections caused by *Campylobacter* spps. *Salmonella* spps. (https://www.fao.org/3/y5159e/y5159e08.htm).

4.4 Antibiotics common in human and animal health

Certain antibiotics are exclusively used in humans, such as isoniazid, to treat tuberculosis; some are used solely in food animals like ionophores and flavonophospholipids. But most antibiotics that are vital for human health are also used in food animals for prophylaxis, treatment, and growth promotion [54]. Cephalosporins are one of the most widely used bactericidal antibiotics that comprise approximately 50-70 percent of all antibiotics used by humans. As per reports, it is one of the top 5 most prescribed antibiotics in Europe [41]. Its mode of action is very similar to that of β-lactam antibiotics as it targets PBPs (penicillin-binding proteins) that play a vital role in cell wall synthesis. Few

TABLE 4.1
Different antibiotics used in food-producing animals and humans if applicable

Class of antibiotic	Mode of action	Name of antibiotic	Use in food-producing animals			Approved for use in humans?	Used in which animals?	Additional comments
			Growth promoter	Prophylaxis	Treatment			
Cephalosporins	Disrupt cell wall synthesis by binding to PBPs (penicillin-binding proteins)	Cephalothin			✓	No	Dairy cattle	-
		Cephalexin			✓	Yes	Dairy cattle	-
		Cefazolin			✓	Yes	Dairy cattle	-
		Cephapirin			✓	Yes	Dairy cattle	-
		Cefuroxime			✓	Yes	Dairy cattle	-
		Ceftiofur		✓		No	Dairy cattle, horses, goats, poultry	-
		Cefiquinome			✓	No	Dairy cattle, pigs	-
Cyclic peptide	Inhibition of cell wall synthesis	Bacitracin	✓	✓	✓	Yes	Cattle, poultry, pigs	-
Glycolipid	Inhibition of cell wall synthesis	Bambermycin	✓	✓		No	Cattle, poultry, pigs	-
Qunioxaline	Inhibition of DNA synthesis	Carbadox	✓	✓		No	Pigs	Banned in 1999 in EC (European Committee) as a growth promoter.
Polymyxin	Binds to LPS (liposaccharides) and disrupts the cell membrane	Colistin		✓	✓	Yes	Pigs, poultry, cows	Banned in 2017 in China as a growth promoter.
Glycopeptide	Inhibition of cell wall synthesis	Avoparcin	✓			No	Pigs, poultry	Banned in 1997 in EC as a growth promoter.
Beta-lactams	Disrupt cell wall synthesis by binding to PBPs (penicillin-binding proteins)	Penicillin	✓		✓	Yes	Cattle, poultry, pigs	Banned in 1972-74 in EC (European Committee) as a growth promoter.
		Ampicillin			✓	Yes	Pigs	-
		Amoxicillin			✓	Yes		-
Tetracycline	Inhibition of protein synthesis by binding to 30S ribosomal subunit	Chlortetracycline	✓	✓		Yes	Cattle, poultry, pigs	Banned in 1972-74 in EC (European Committee) as a growth promoter.
		Oxytetracycline	✓	✓	✓	Yes	Cattle, fish	
Aminoglycosides	Inhibition of protein synthesis by binding to 30S ribosomal subunit	Gentamicin			✓	Yes	Cattle, poultry, pigs	-
					✓	Yes	Poultry	-
		Neomycin			✓	Yes	Poultry	Banned in 1972-74 in EC (European Committee) as a growth promoter.
Macrolides	Inhibition of protein synthesis by binding to 50S ribosomal subunit	Erythromycin	✓	✓	✓	Yes	Poultry, pigs	-
		Tylosin	✓	✓	✓	No	Cattle, poultry, pigs	Banned in 1999 in EC as a growth promoter.
		Virginiamycin	✓	✓	✓	No	Poultry, pigs	Banned in 1998 and 1999 in Denmark and EC, respectively, as a growth promoter.

cephalosporins are currently present exclusively for veterinary use, such as cephalothin, ceftiofur, and cefquinome. There are many first-generation cephalosporins such as cephalexin, cefazolin, and cephalosporin; second-generation cephalosporins such as cefuroxime and third-generation cephalosporins such as cefoperazone that humans use that are also approved for use in food animals. These drugs are generally used to treat bacterial infections in animals, such as mastitis in lactating cows [55]. In 2008, the Food and Drug Administration (FDA) announced a rule that banned "extra-label use" of cephalosporins in animals. However, this rule was withdrawn after several veterinarians and pharmaceutical companies protested against it and later revised again in 2012. The new law stated that the "off-label use" of ceftiofur, a third-generation cephalosporin, would be forbidden in food-producing animals. This action was taken after it was observed that large doses of this drug were frequently given to dairy cattle before they were slaughtered. Injections were regularly administered into poultry eggs in hatcheries. One of the consequences of this was a steep rise in ceftiofur resistant *Salmonella enterica* serovar Heidelberg samples from animal products over the years. This was a matter of concern since pathogens resistant to ceftiofur will mostly gain resistance to other cephalosporins routinely used to treat infections in humans [19]. Similar actions were taken by Japan and Canada that restricted its prophylaxis use in 2012 and 2014, respectively [55]. In 2010-13, a positive outcome was seen wherein there was a significant decrease in cephalosporin-resistant *E. coli* isolates from broilers in Japan [56]. A study was conducted in the mid-1990s that determined that the frequent usage of avoparcin, a glycopeptide exclusively used in animals as a growth promoter, swine, and poultry, led to an increased occurrence of VREF (vancomycin resistance *E. faecium*). Vancomycin also belongs to the glycopeptide class of antibiotics. It is the first choice for treating severe Gram-positive infections such as MRSA (methicillin-resistant *S. aureus*) [29]. Colistin, belonging to the polymyxin class of antibiotics, is used as a last-resort drug to treat severe Gram-negative infections such as pulmonary exacerbations due to MDR (multiple drug resistant) *P.aeruginosa* in cystic fibrosis patients. Even though its side-effects involve high levels of nephrotoxicity and neurotoxicity in humans, it is shown to be highly effective, which is why it is still being used [41]. As per ECDC (European Center for Disease Prevention and Control) reports in 2015, it was used in alarmingly large quantities in food animals such as swine, cattle, and poultry for prophylaxis, metaphylaxis, and treatment purposes, mainly of *Enterobacteriaceae* infections in China, Europe, and Brazil [48]. A plasmid resistance gene, mcr-1, encoding for phosphoethanolamine transferase was detected in *E. coli* samples collected from swine, cattle, and poultry products from China. This enzyme catalyses the addition of phosphoethanolamine on LPS (lipopolysaccharide) of the bacterial cell membrane, thus, changing its overall charge. This gene confers colistin resistance to the bacteria[55]. This gene can also be transmitted via conjugation to other bacteria such as *K. pnuemoniae, P. aeruginosa,* etc. This gene was also found in human blood and environmental water samples. Newer variants of this gene are also reported [55]. China itself was utilising 8000 tonnes of colistin annually out of the overall 12000 tonnes of colistin produced worldwide. This was one of the primary reasons behind increased colistin resistance in humans and food animals in China. To curb this problem, the Chinese government declared a total ban on colistin as an additive in animal feed in 2016 as shown in Figure 4.2. Following this, FSSAI (Food Safety and Standards Authority of India) announced a ban on colistin use [42]. The above case studies demonstrate how increasing AMR in food-producing animals impacts humans.

*In 2007, majority of all AGPs were banned in Mexico except avilamycin, avoparcin, bacitracin, bambermycin, monensin, salinomycin, spiramycin, tylosin, vancomycin, and virginiamycin.

4.5 Future perspectives

Several complicated mechanisms contribute to the global spread of AMR. More research is needed to understand how common antimicrobials used in both people and animals might directly or indirectly influence human health. When taking steps to minimise the spread of AMR, it is critical to consider all of the other crucial

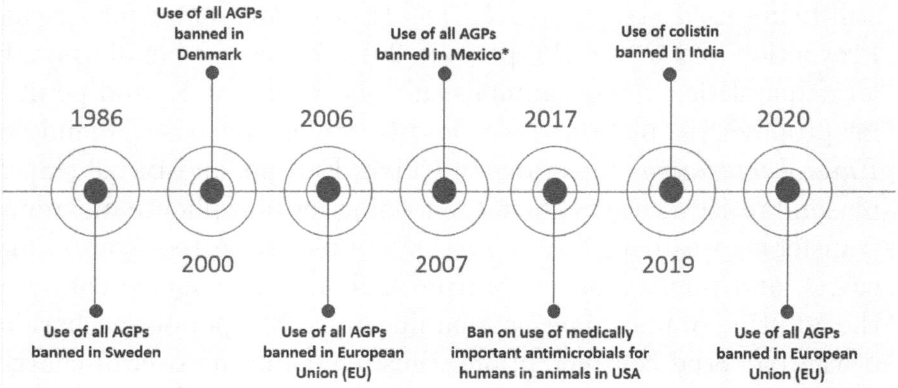

FIGURE 4.2
Timeline of the ban of AGP across the globe (Ma et al., 2021).

factors. Imposing a complete prohibition on the use of antimicrobials as growth boosters in food-producing animals, as China did in 2020 [57], seems to be a good option temporarily; however, in the long run, this can create a lot of problems. As per a WHO report in 2003, there was an increase in death rate by 0.6 percent, a 2.6 percent drop in growth rate and about a 2 percent decrease in feed efficiency in pigs after the ban of AGPs in Denmark. An economic loss of about 10-20 billion USD was further projected [48]. To curb this, investments must be made to find efficient alternatives for antimicrobials in livestock. There are two ways in which the AMR spread can be steepened, first by reducing the antimicrobial dosage given to humans and animals plus limiting the use of last resort; medically essential antimicrobials and broad-spectrum drugs and secondly; by prohibiting or pre-treating the waste before dumping it into the environment which would help reduction of antimicrobial residues in foods and water [58]. WHO guidelines recommend reduced use of medically-important antimicrobials in food-producing animals for treatment and complete restriction of its use for prophylaxis and growth promotion. Research data and recommendations from interdisciplinary sources can help WHO reduce the use of antimicrobials. CRISPR/Cas 9 (clustered regularly interspaced short palindromic repeats- associated protein 9) is a popular genome editing system that can be used to specifically target antibiotic-resistant pathogens without harming the normal microflora. This technology was used to target and cleave out antibiotic resistance genes in the plasmid of ES(extended-spectrum β-lactamases) producing *E. coli* (multiple drug-resistant bacteria) which caused it to revert to its original sensitive state [59]. Another potential substitute is the use of bacteriophages. Studies were conducted to understand if phage therapy could be utilised to control the spread of livestock-associated pathogens, e.g., *E. coli* O157:H7 in ruminants [60] vancomycin-resistant *Enterococcus faecium* [61], etc. Nonantibiotic growth promoters such as enzymes in feed, competitive exclusion products, prebiotics and probiotics can be used as alternatives. Other animal health treatments that are underused in underdeveloped nations, including vaccines, vector control, disinfectants, phytotherapy, and phage therapy can be improved [25]. To combat the global matter of AMR, the realisation of this problem should be interdisciplinary, including medicine, agriculture, wastewater, food industry, etc. It would also require international agencies like WHO, FAO, and UN to intervene regularly. Not only these, but research data from other disciplines like health economics, social science, and international law would be additionally required. These can be done by supporting collaborative action of all sectors, including producers, consumers, and mediators, via mutual and 'One Health' awareness of the concerns and educated solutions through antibiotic stewardship.. Moreover, the antimicrobial drug pipeline needs to be strengthened to provide alternative approaches for decreasing this problem [62].

Bibliography

[1] Gina Devasahayam, William M Scheld, and Paul S Hoffman. Newer antibacterial drugs for a new century. *Expert opinion on investigational drugs*, 19(2):215–234, 2010.

[2] Alison H Holmes, Luke SP Moore, Arnfinn Sundsfjord, Martin Steinbakk, Sadie Regmi, Abhilasha Karkey, Philippe J Guerin, and Laura JV Piddock. Understanding the mechanisms and drivers of antimicrobial resistance. *The Lancet*, 387(10014):176–187, 2016.

[3] Rustam I Aminov. A brief history of the antibiotic era: lessons learned and challenges for the future. *Frontiers in microbiology*, 1:134, 2010.

[4] Porooshat Dadgostar. Antimicrobial resistance: implications and costs. *Infection and drug resistance*, 12:3903, 2019.

[5] Longzhu Cui, Xiaoxue Ma, Katsuhiro Sato, Keiko Okuma, Fred C Tenover, Elsa M Mamizuka, Curtis G Gemmell, Mi-Na Kim, Marie-Cecile Ploy, N El Solh, et al. Cell wall thickening is a common feature of vancomycin resistance in staphylococcus aureus. *Journal of clinical microbiology*, 41(1):5–14, 2003.

[6] Keiichi Hiramatsu. Vancomycin-resistant staphylococcus aureus: a new model of antibiotic resistance. *The Lancet infectious diseases*, 1(3):147–155, 2001.

[7] Haley J Morrill, Jason M Pogue, Keith S Kaye, and Kerry L LaPlante. Treatment options for carbapenem-resistant enterobacteriaceae infections. In *Open forum infectious diseases*, volume 2. Oxford University Press, 2015.

[8] Vincent H Tam, Cary A Rogers, Kai-Tai Chang, Jaye S Weston, Juan-Pablo Caeiro, and Kevin W Garey. Impact of multidrug-resistant pseudomonas aeruginosa bacteremia on patient outcomes. *Antimicrobial agents and chemotherapy*, 54(9):3717–3722, 2010.

[9] Richard H Gustafson and RE Bowen. Antibiotic use in animal agriculture. *Journal of applied microbiology*, 83(5):531–541, 1997.

[10] Abigail A Salyers. *Antibiotic resistance gene transfer in the mammalian intestinal tract: implications for human health, food safety and biotechnology*. RG Landes Company, 1995.

[11] Yi-Yun Liu, Yang Wang, Timothy R Walsh, Ling-Xian Yi, Rong Zhang, James Spencer, Yohei Doi, Guobao Tian, Baolei Dong, Xianhui Huang, et al. Emergence of plasmid-mediated colistin resistance mechanism mcr-1 in animals and human beings in china: a microbiological and molecular biological study. *The Lancet infectious diseases*, 16(2):161–168, 2016.

[12] Scott A McEwen and Paula J Fedorka-Cray. Antimicrobial use and resistance in animals. *Clinical infectious diseases*, 34(Supplement_3):S93–S106, 2002.

[13] David W Nelson, John E Moore, and Juluri R Rao. Antimicrobial resistance (amr): significance to food quality and safety. *Food quality and safety*, 3(1):15–22, 2019.

[14] Michael Marmot and Ruth Bell. The sustainable development goals and health equity. *Epidemiology*, 29(1):5–7, 2018.

[15] Brad Spellberg, Gail R Hansen, Avinash Kar, Carmen D Cordova, Lance B Price, and James R Johnson. Antibiotic resistance in humans and animals. *NAM Perspectives*, 2016.

[16] Stephanie Fletcher. Understanding the contribution of environmental factors in the spread of antimicrobial resistance. *Environmental health and preventive medicine*, 20(4):243–252, 2015.

[17] Thomas P Van Boeckel, Charles Brower, Marius Gilbert, Bryan T Grenfell, Simon A Levin, Timothy P Robinson, Aude Teillant, and Ramanan Laxminarayan. Global trends in antimicrobial use in food animals. *Proceedings of the National Academy of Sciences*, 112(18):5649–5654, 2015.

[18] C Lee Ventola. The antibiotic resistance crisis: part 1: causes and threats. *Pharmacy and therapeutics*, 40(4):277, 2015.

[19] Sally R Partridge, Stephen M Kwong, Neville Firth, and Slade O Jensen. Mobile genetic elements associated with antimicrobial resistance. *Clinical microbiology reviews*, 31(4):e00088–17, 2018.

[20] Wanda Reygaert. Methicillin-resistant staphylococcus aureus (mrsa): molecular aspects of antimicrobial resistance and virulence. *Clinical Laboratory Science*, 22(2):115, 2009.

[21] GIUSEPPE CORNAGLIA, ANNARITA MAZZARIOL, ROBERTA FONTANA, and GIUSEPPE SATTA. Diffusion of carbapenems through the outer membrane of enterobacteriaceae and correlation of their activities with their periplasmic concentrations. *Microbial drug resistance*, 2(2):273–276, 1996.

[22] Jessica Blair, Mark A Webber, Alison J Baylay, David O Ogbolu, and Laura JV Piddock. Molecular mechanisms of antibiotic resistance. *Nature reviews microbiology*, 13(1):42–51, 2015.

[23] Lokender Kumar, Sanjay Chhibber, and Kusum Harjai. Zingerone inhibit biofilm formation and improve antibiofilm efficacy of ciprofloxacin against pseudomonas aeruginosa pao1. *Fitoterapia*, 90:73–78, 2013.

[24] Andrew F Read and Robert J Woods. Antibiotic resistance management. *Evolution, medicine, and public health*, 2014(1):147, 2014.

[25] Bogdan-Ioan Coculescu. Antimicrobial resistance induced by genetic changes. *Journal of medicine and life*, 2(2):114, 2009.

[26] Stig Wall. Prevention of antibiotic resistance–an epidemiological scoping review to identify research categories and knowledge gaps. *Global Health Action*, 12(sup1):1756191, 2019.

[27] Aastha Chokshi, Ziad Sifri, David Cennimo, and Helen Horng. Global contributors to antibiotic resistance. *Journal of global infectious diseases*, 11(1):36, 2019.

[28] Qiuzhi Chang, Weike Wang, Gili Regev-Yochay, Marc Lipsitch, and William P Hanage. Antibiotics in agriculture and the risk to human health: how worried should we be? *Evolutionary applications*, 8(3):240–247, 2015.

[29] Antonet Svircev, Dwayne Roach, and Alan Castle. Framing the future with bacteriophages in agriculture. *Viruses*, 10(5):218, 2018.

[30] Angkana Lekagul, Viroj Tangcharoensathien, and Shunmay Yeung. Patterns of antibiotic use in global pig production: a systematic review. *Veterinary and animal science*, 7:100058, 2019.

[31] Ilse Overdevest, Ina Willemsen, Martine Rijnsburger, Andrew Eustace, Li Xu, Peter Hawkey, Max Heck, Paul Savelkoul, Christina Vandenbroucke-Grauls, Kim van der Zwaluw, et al. Extended-spectrum β-lactamase genes of escherichia coli in chicken meat and humans, the netherlands. *Emerging infectious diseases*, 17(7):1216, 2011.

[32] Delia Grace. Review of evidence on antimicrobial resistance and animal agriculture in developing countries. 2015.

[33] Isabel Frost, Thomas P Van Boeckel, João Pires, Jessica Craig, and Ramanan Laxminarayan. Global geographic trends in antimicrobial resistance: the role of international travel. *Journal of travel medicine*, 26(8):taz036, 2019.

[34] Panagiotis Efthymiou, Despoina Gkentzi, and Gabriel Dimitriou. Knowledge, attitudes and perceptions of medical students on antimicrobial stewardship. *Antibiotics*, 9(11):821, 2020.

[35] Subhash C Arya. Quality control of essential drugs. *The Lancet*, 350(9084):1106–1107, 1997.

[36] Lars Bogø Jensen, Anette M Hammerum, Frank M Aarestrup, Anthony E van den Bogaard, and Ellen E Stobberingh. Occurrence of sata and vgb genes in streptogramin-resistant enterococcus faecium isolates of animal and human origins in the netherlands. *Antimicrobial Agents and Chemotherapy*, 42(12):3330–3331, 1998.

[37] Britta Ballhausen, Philipp Jung, André Kriegeskorte, Phuti Edward Makgotlho, Ulla Ruffing, Lutz von Müller, Robin Köck, Georg Peters, Mathias Herrmann, Wilma Ziebuhr, et al. La-mrsa cc398 differ from classical community acquired-mrsa and hospital acquired-mrsa lineages: functional analysis of infection and colonization processes. *International Journal of Medical Microbiology*, 304(7):777–786, 2014.

[38] David JP Mallon, John E Corkill, Sarah M Hazel, J Sian Wilson, Nigel P French, Malcolm Bennett, and C Anthony Hart. Excretion of vancomycin-resistant enterococci by wild mammals. *Emerging infectious diseases*, 8(6):636, 2002.

[39] Luc A Devriese, Margaretha Ieven, Herman Goossens, Peter Vandamme, Bruno Pot, Jozef Hommez, and Freddy Haesebrouck. Presence of vancomycin-resistant enterococci in farm and pet animals. *Antimicrobial Agents and Chemotherapy*, 40(10):2285–2287, 1996.

[40] Ian Phillips, Anna King, Bernard Rowe, Susannah Eykyn, WR Gransden, JenniferA Frost, and RogerJ Gross. Epidemic multiresistant escherichia coli infection in west lambeth health district. *The Lancet*, 331(8593):1038–1041, 1988.

[41] Sandra I Berríos-Torres, Craig A Umscheid, Dale W Bratzler, Brian Leas, Erin C Stone, Rachel R Kelz, Caroline E Reinke, Sherry Morgan, Joseph S Solomkin, John E Mazuski, et al. Centers for disease control and prevention guideline for the prevention of surgical site infection, 2017. *JAMA surgery*, 152(8):784–791, 2017.

[42] Massimo Sartelli, Timothy C Hardcastle, Fausto Catena, Alain Chichom-Mefire, Federico Coccolini, Sameer Dhingra, Mainul Haque, Adrien Hodonou, Katia Iskandar, Francesco M Labricciosa, et al. Antibiotic use in low and middle-income countries and the challenges of antimicrobial resistance in surgery. *Antibiotics*, 9(8):497, 2020.

[43] Ramanan Laxminarayan, Adriano Duse, Chand Wattal, Anita KM Zaidi, Heiman FL Wertheim, Nithima Sumpradit, Erika Vlieghe, Gabriel Levy Hara, Ian M Gould, Herman Goossens, et al. Antibiotic resistance—the need for global solutions. *The Lancet infectious diseases*, 13(12):1057–1098, 2013.

[44] Ian Phillips, Mark Casewell, Tony Cox, Brad De Groot, Christian Friis, Ron Jones, Charles Nightingale, Rodney Preston, and John Waddell. Does the use of antibiotics in food animals pose a risk to human health? a critical review of published data. *Journal of antimicrobial chemotherapy*, 53(1):28–52, 2004.

[45] B Vaseeharan, P Ramasamy, T Murugan, and JC Chen. In vitro susceptibility of antibiotics against vibrio spp. and aeromonas spp. isolated from penaeus monodon hatcheries and ponds. *International journal of antimicrobial agents*, 26(4):285–291, 2005.

[46] Tara C Smith, Wondwossen A Gebreyes, Melanie J Abley, Abby L Harper, Brett M Forshey, Michael J Male, H Wayne Martin, Bayleyegn Z Molla, Srinand Sreevatsan, Siddhartha Thakur, et al. Methicillin-resistant staphylococcus aureus in pigs and farm workers on conventional and antibiotic-free swine farms in the usa. *PLoS One*, 8(5):e63704, 2013.

[47] Jayaseelan Murugaiyan, P Anand Kumar, G Srinivasa Rao, Katia Iskandar, Stephen Hawser, John P Hays, Yara Mohsen, Saranya Adukkadukkam, Wireko Andrew Awuah, Ruiz Alvarez Maria Jose, et al. Progress in alternative strategies to combat antimicrobial resistance: Focus on antibiotics. *Antibiotics*, 11(2):200, 2022.

[48] European Antimicrobial Resistance Surveillance System (EARSS et al. Summary of latest data on antibiotic resistance in the european union. *http://www. ecdc. europa. eu/en/activities/surveillance/EARS-Net/Pages/index. aspx*, 2010.

[49] Gabriel G Perron, Lyle Whyte, Peter J Turnbaugh, Jacqueline Goordial, William P Hanage, Gautam Dantas, and Michael M Desai. Functional characterization of bacteria isolated from ancient arctic soil exposes diverse resistance mechanisms to modern antibiotics. *PloS one*, 10(3):e0069533, 2015.

[50] JIR Castanon. History of the use of antibiotic as growth promoters in european poultry feeds. *Poultry science*, 86(11):2466–2471, 2007.

[51] André G Buret. Immuno-modulation and anti-inflammatory benefits of antibiotics: the example of tilmicosin. *Canadian Journal of Veterinary Research*, 74(1):1–10, 2010.

[52] Mee-Ngan F Yap. The double life of antibiotics. *Missouri medicine*, 110(4):320, 2013.

[53] Bonnie M Marshall and Stuart B Levy. Food animals and antimicrobials: impacts on human health. *Clinical microbiology reviews*, 24(4):718–733, 2011.

[54] Scott A McEwen and Peter J Collignon. Antimicrobial resistance: a one health perspective. *Microbiology spectrum*, 6(2):6–2, 2018.

[55] Pauline Hood and Ehsan Khan. *Understanding Pharmacology in Nursing Practice*. Springer, 2020.

[56] John S Mackenzie, Martyn Jeggo, Peter Daszak, and Juergen A Richt. *One Health: The human-animal-environment interfaces in emerging infectious diseases*, volume 366. Springer, 2013.

[57] Feiyang Ma, Shixin Xu, Zhaoxin Tang, Zekun Li, and

Lu Zhang. Use of antimicrobials in food animals and impact of transmission of antimicrobial resistance on humans. *Biosafety and Health*, 3(01):32–38, 2021.

[58] Wenguang Xiong, Yongxue Sun, and Zhenling Zeng. Antimicrobial use and antimicrobial resistance in food animals. *Environmental Science and Pollution Research*, 25(19):18377–18384, 2018.

[59] Jun-Seob Kim, Da-Hyeong Cho, Myeongseo Park, Woo-Jae Chung, Dongwoo Shin, Kwan Soo Ko, and Dae-Hyuk Kweon. Crispr/cas9-mediated re-sensitization of antibiotic-resistant escherichia coli harboring extended-spectrum β-lactamases. *Journal of Microbiology and Biotechnology*, 26(2):394–401, 2016.

[60] Gong Cheng, Yongfei Hu, Yeshi Yin, Xi Yang, Chunsheng Xiang, Baohong Wang, Yanfei Chen, Fengling Yang, Fang Lei, Na Wu, et al. Functional screening of antibiotic resistance genes from human gut microbiota reveals a novel gene fusion. *FEMS microbiology letters*, 336(1):11–16, 2012.

[61] Biswajit Biswas, Sankar Adhya, Paul Washart, Brian Paul, Andrei N Trostel, Bradford Powell, Richard Carlton, and Carl R Merril. Bacteriophage therapy rescues mice bacteremic from a clinical isolate of vancomycin-resistant enterococcus faecium. *Infection and immunity*, 70(1):204–210, 2002.

[62] Mark Woolhouse, Melissa Ward, Bram Van Bunnik, and Jeremy Farrar. Antimicrobial resistance in humans, livestock and the wider environment. *Philosophical Transactions of the Royal Society B: Biological Sciences*, 370(1670):20140083, 2015.

5

Antimicrobial and Antibiotic Resistance in Developing Countries: Health Economics, Global Governance, and Sustainable Development Goals

Stefano Greco
Chitkara Spark Centre for Multidisciplinary European Studies, Chitkara University Punjab, India

Romans Putans
Department of International Business and Economics, Riga Stradins University, Latvia

Lauma Springe
Department of Public Health and Epidemiology, Riga Stradins University, Latvia

CONTENTS

5.1 Introduction

The discovery of antibiotics and the developments of modern medicine are universally included among the greatest achievements of mankind. Unfortunately, a set of natural and man-made dynamics is compromising the efficacy of the drugs regularly utilised to cure patients affected by communicable diseases. As in the case of climate change, human activities are accelerating natural processes. Since the first stage of life on Earth, bacteria mutated as a biological response to the surrounding environment. In contemporary times, this natural process of evolution has accelerated due to the misuse of antibiotics in human and animal health, the pharmaceutical contamination of water and soil; as well as following the adoption of a governance model leaving the development of new pharmaceutical products conditional to the economic assessment of private stakeholders. Although traditional and digital media are hardly informing society over the causes and the impact of antimicrobial resistance, a consolidated cluster of studies demonstrated that in the long-term drug-resistant infections could become the first cause of death both in developed nations and in low- and middle-income countries (LMICs).

Formalized in 2015 by the UN Assembly with the unanimous approval of the resolution *Transforming our world: the 2030 Agenda for Sustainable Development*, the Sustainable Development Goals (SDGs) represent the major attempt to enforce a global approach promoting human-centric progress . Composed of 17 goals and 169 targets, the *2030 Agenda* establishes an integrated action plan. Built upon the leave-no-one behind the principle, the SDGs comprehensive strategy is based on interconnected social, economic, and environmental pillars, designed to improve the effectiveness of international, national, and regional development instruments. The global commitment is by 2030 to deliver coordinated actions tackling socio-economic inequalities, promoting a more balanced relationship between human progress, economic growth, and environmental preservation. As part of this integrated strategy, composed of 13 intertwined targets, the third SDG Good Health and Well-being aims to foster coordinated global actions improving people's lives through the development of public healthcare infrastructures, services, and know-how. At the same time, excluded from the outreach of the 2030 Agenda, antimicrobial resistance (AMR) creates reasoned concerns over the effectiveness of global healthcare investments. The concerns are mainly related to the health capabilities and the vast inequalities in terms of access to medical services observed in the LMICs.

The chapter aims to assess how the projected explosion of drug-resistant infections is expected to affect the achievement of the SDGs in developing countries. As a preliminary step, the chapter presents a detailed overview of the direct and indirect socio-economic costs resulting from the emergence of drug-resistant diseases, intersected with a critical assessment of the SDGs designed to improve healthcare services and societal well-being. Crossing the key findings related to the socio-economic impact of AMR and the UN developmental goal in the field of health, the final section assesses whether in developing countries the rising frequency of AMR is likely to impact the capabilities of national authorities in achieving the SDGs targets and expand healthcare services.

5.2 The economic impact of antimicrobial and antibiotic resistance

Although the impact of AMR and antibiotic resistance (ABR) is largely analysed restricting the assessment to the health and social domains, the exponential growth of pathogens resistant to drugs have relevant economic implications. Drug-resistant infections significantly inflated the cost of medical treatment mainly due to a surge in hospitalisation length, increased number of professional consultations required by the patients, as well as due to the rising number of patients requiring hospitalisation. The economic impact is not limited to the healthcare level. In fact, along with the supplementary expenditure related to hospitalisation and health services, the effect of antimicrobial resistance entangles additional costs at the macro and microeconomic levels [1].

With a combined cost of up to 100 trillion to the global economy – pushing 28 million people into extreme poverty [2] – the rise of bacteria, viruses, fungi and parasites resistant to existing medications represent one of the most pressing global socio-economic challenges. Most troublingly, if policymakers will fail in adopting policies diminishing the progression of AMR, there is the risk to invalidate most of the common procedures and treatments developed by modern medicine [3]. The World Bank's 2019 report, Pulling Together to Beat Superbugs, found that AMR already causes 700 000 deaths per year. Without tangible actions, the death toll from AMR could rise even higher. By 2050, AMR and ABR could cause as many as 10 million victims per annum [4]. By the same year, it has been also calculated that the casualties related to AMR would diminish the world estimated population from a range of 11 million to 444 million people [5].

While reliable data are hard to find, several studies have tried to quantify the cumulative economic impact of drug-resistant conditions. The World Bank (2017) estimated that the AMR by 2050 could generate a net loss in the global GDP equivalent to 2.5 percent, combined with the increment of extreme poverty primarily in developing countries and a relevant shrink of the world trade volume and livestock production. In the most advanced economies, the additional cost resulting from the need to provide supplementary treatments is expected to cost the public healthcare scheme an average of 25 000 per patient affected by drug-resistant pathogens [6]. The European Centre for Disease Control and Prevention (2009) has estimated that every year, in Europe alone, AMR is responsible for approximately 25 000 deaths, causing additional investments in the healthcare systems for at least €1.5 billion. A similar figure has been projected for North America. Every year, the US Centers for Disease Control and Prevention (2013) estimate AMR and ABR to cause 23 000 deaths, 20 billion additional expenditure related to supplementary health services, and 35 billion losses due to the decrement of productivity in the US labour force.

The global increase in quantitative and geographical terms of antibiotic consumption between 2000 and 2015 is undoubtedly noticeable [7]. Another study from 2019 conducted by US Centers for Disease Control and Prevention on antibiotics use in the United

States show that at least 30 percent (or about 47 million antibiotic courses) each year are prescribed for conditions that do not need antibiotics. Despite the absence of clear data, overuse and overprescription of antibiotics are an additional factor directly – and even more, indirectly – enlarging the economic impact of antimicrobial resistance. Similarly to human health, the overconsumption of antibiotics in livestock – and the subsequent emergence of AMR in animal diseases – has direct and indirect financial consequences. The majority of antibiotics consumed worldwide are utilised for intensive animal farming [8]. Despite the magnitude of the problem, data aggregating the cumulative global financial impact on livestock is missing. Nevertheless, several studies quantified the financial losses generated by drug-resistant pathogens in intensive animal agriculture. For instance, in six years, aquafarming industries based in Thailand, China, Malaysia, Mexico, and Thailand faced a loss of 44 billion due to resistant infections [9]. Similar losses are experienced also by entrepreneurs working with intensive farming on the land. Calculating the AMR economic impact from another perspective, a team of scientists from the Johns Hopkins University created a mathematical model that estimated a yearly loss of at least 33 million for the US society imputable to foodborne diseases caused by the consumption mainly of poultry contaminated with resistant bacteria [10].

Particularly in low-income countries, it is still a challenge to quantify the economic impact of AMR and ABR. This is due to the complexity and variability of the impact factors, as well as to the complexity to detect (and quantify) cause-and-effect dynamics. As a result, there is a dearth of reliable data estimating the economic impact of the AMR in LMICs. In a recent international joint review [11] scientists from the United Kingdom, Hong Kong and Singapore identified only 110 articles from 1998 to 2019 aiming at quantifying AMR costs. The authors themselves remarked on the necessity to develop better methods to eliminate biases and spurious correlations. Investigating the availability and use of antibiotics in 20 LMICs, another recent research [12] highlighted the need to improve the data collection practices in developing and transition economies. At the same time, the situation is gradually improving as a higher number of countries are sharing standardised data in the framework of global overview coordinated by intergovernmental organisations. For instance, the number of countries participating in the survey conducted by the World Organization for Animal Health (derived organisation under UN) focusing on antimicrobial agents intended for use in animals, have increased from 130 countries in 2015 to 155 countries in 2017 [13]. Regarding this matter, the ability of a country to provide information reflects not only its political readiness for global engagement but mostly the growing domestic capacity in collecting detailed data on antimicrobial agents.

Despite the global nature of the phenomenon, the economic impact of the AMR is higher in LMICs. For instance, while the global GDP is expected to slow down by 2.5 percent in the next 35 years, during the same period, the GDP of the developing countries is expected to decline by 5 percent [14]. At the same time, while in North America and Europe 17 percent of the analysed pathogens are resistant to treatments, half of the bacterial infections occurring in Russia, India, and Brazil are hardly treatable with the available medicines [15]. Antimicrobial resistance is also expected to exacerbate inequalities. Even though in absolute terms the economic impact on the public healthcare sector is lower for the fact that developing countries scarcely invest in health infrastructures and services, the increasing infection morbidity is likely to increase the frequency of out-of-pocket medical expenditure for disadvantaged socio-economic individuals [16]. As a result, in countries that do not offer affordable public health services, marginalised people will have to choose to bear an unsustainable family debt or risk their lives. Along with the increased AMR fatality in developing countries, these dynamics will generate losses in terms of human capital, increasing poverty and decelerating the socio-economic growth of the LMICs. Feeding inequality and poverty, the exponential growth of AMR will further increase the existing economic distance between rich and poor countries.

5.3 Global governance, health, and Sustainable Development Goals

Launched as the United Nations (UN) flagship initiative, the Sustainable Development Goals (SDGs) represent the most tangible contemporary attempt to transnationally coordinate domestic development policies. Despite the non-legally binding and voluntary attainment of the targets, the SDGs represent a central novelty in the global governance panorama. The uniqueness of the SDGs lies in the wide-ranging visibility acquired by the targets notwithstanding the absence of a dedicated political-bureaucratic apparatus sustaining the program. Indeed, during the three years of negotiation, the definition of the SDGs targets has been strategically detached from the ratification of international treaties, and/or to the establishment of an intergovernmental secretariat centralising the procedures for the grassroots implementation [17]. This intergovernmental approach – based on shared goals, pursued through subsidiarity principle and the local ownership of the initiatives launched – resulted in a non-confrontational and national-driven political process [18]. As a tangible result of the experience matured during the implementation of the Millennium Development Goals, the UN adopted an integrated approach aiming to improve the coordination, the comprehensives (and ultimately the impact) of the international and national policies addressing social and human development.

In the field of health, the passage from the eight Millennium Development Goals to the 17 SDGs has been ultimately characterised by the adoption of a horizontal intersectoral system, replacing a vertical silos-based approach drafting uncoordinated strategies tackling single diseases [19]. Arguments advocating the effectiveness of open global governance built upon non-legally binding goals should be assessed considering the contemporary absence of viable alternatives [2]. From a realistic point of view, the emergence of a chaotic and liquid multipolarity has paralysed the governance of the key global intergovernmental organisations, opening the way for pragmatic and flexible coordination models. Hence, while the achievement of the SDGs primarily remains a national affair, at the same time an open method of coordination represents the sole short-term solution to mediate between opposing domestic positions. Grounded on an extensive set of targets, the flexible model adopted by the UN in the 2030 Agenda resolution is meant to ease the mainstreaming and the adoption of the SDGs by national governments, creating favourable pre-conditions for the achievements of the envisaged results [20]. Responding to critics related to the one-fits-all and post-colonial approach raised by the Millennium Development Goals [21, 22], the SDGs do not separate countries according to their developmental stage. In other words, the country with the highest GDP per capita shares the same goals and targets with the political community having the lowest Human Development Index. Moreover, local ownership is favoured by the flexibility and the adaptability of the tools that domestic political and institutional players can deploy to achieve the expected results. In this scenario, the success of the SDGs depends upon the effective translation of global aspiration into national commitment [23].

Considering the background and the current challenges faced by multilateralism, the 2030 Agenda and the SDGs represent the most comprehensive global initiative to globally improve the healthcare systems. In contrast with the Millennium Development Goals strategy, the SDGs respond to the necessity to reinforce national health capabilities by enhancing the ability to deliver services, at the same time, to promote investment fostering the sustainability of the healthcare infrastructures [24].

There exist multiple mutual interdependencies between the state of art in the healthcare sector and the improvements of the social, economic, and environmental domains. The state of the art in the healthcare sector and developments in the social, economic, and environmental sectors have various mutual interdependencies. As highlighted in the quantitative investigation conducted by [25], the investments aiming to achieve the goals in the field of public health are also instrumental to the success of development policies targeting domains not related to health. For example, improving the accessibility to health services would reduce inequalities and poverty, and at the same time, it would enhance economic growth, educational level, and gender equality [26]. Simultaneously, the success of the health-related SDGs is conditional to tangible progress in the socio-economic and environmental domains [27]. For instance, economic policies increasing individual and family income, and concrete actions in favour of gender equality, would tackle malnutrition (SDG 2.2), while also improving women's health (SDG 5.6). Similarly, concrete achievement in the environmental field – minimising the contamination of air, water, and soil (SDG 12.5) – would automatically result in a global reduction of deaths caused by non-communicable diseases. In a nutshell, the achievement of the health targets is required for the social-economic success of the SDGs, as much tangible socio-economic and environmental progress are essential to meet the targets and indicators listed under the SDG 3 Good Health and Well-being (Figure 5.1.).

Restricting the analysis to the SDGs targets entirely focusing on healthcare systems and the improvement of the well-being in developing countries, the envisaged emergence of universal health coverage schemes (SDG 3.8) represents the most ambitious – and at the same time pivotal – health target set by the 2030 Agenda. The SDG 3 designs a future where all the individuals would have free access to quality health services, tackling the health inequalities currently existing in most of the UN member states [28, 29].

The inclusion of the universal health coverage (UHC) within the SDGs targets can be related to the reiterated endorsements operated by the WHO. In 2014 in a technical report on sustainable health financing and insurance packages, the WHO explicitly encouraged national governments to allocate public resources to ensure universal access to preventive, curative, and rehabilitative health services. Similarly, in the introduction of another report, the experts from the WHO (2010) related the imperative necessity to develop accessible public healthcare facilities with the commitment taken by the member states under the Alma-Ata Declaration to invest in public health as a means for societal development [30].

Mostly for the transitioning and developing countries, the establishment of a UHC regime appears as a preliminary action for the achievement of the SDGs targets related to global health and inequalities [31]. The UN endorsement in favour of the development of public UHC schemes, and the subsequent global mis-

TABLE 5.1
SDGs targets related to health
Source: authors elaboration based on United Nations (2015:17-28)

Targets entirely focusing on health	
SDG 3.1	Reduce global maternal mortality ratio to less than 70 per 100 000 live births
SDG 3.2	Reduce neonatal mortality as low as 12 per 1 000 live births, and for the under-5 mortality to less than 25 per 1 000 live births
SDG 3.3	End epidemics of AIDS, tuberculosis, malaria and forgotten tropical diseases, tackling hepatitis, water-borne diseases, and other communicable diseases
SDG 3.4	Reduce by one third the mortality of non-communicable diseases, promoting mental health and well-being
SDG 3.5	Strengthen the prevention and the treatment of people affected by drugs and alcohol abuse
SDG 3.6	Halve casualties and injuries from road traffic accidents
SDG 3.7	Universal access to sexual and reproductive healthcare services
SDG 3.8	Achieve universal health coverage: financial risk protection, access to health-care services, affordable medicines, and vaccines for all
SDG 3.9	Reduce the casualties and illnesses related to hazardous chemicals and air, water and soil contamination
Targets requiring improvements in national healthcare systems	
SDG 1.4	Ensure that poor and vulnerable citizens have equal access to basic services
SDG 2.2	End all forms of malnutrition, addressing nutritional needs of adolescent girls, pregnant and lactating women, and the elderly.
SDG 5.6	Universal access to sexual and reproductive health
SDG 9.1	Develop reliable, accessible, sustainable, and resilient infrastructure to support human well-being
SDG 10.4	Reduce inequalities through the adoption of social protection policies
SDG 11.5	Significant reduction of deaths involving poor and people in vulnerable situations
SDG 12.5	Significantly reduces the impact on human health derived from the contamination of air, soil, and water

sion to foster accessibility to healthcare services for the socio-economic disadvantaged share of the population, respond to an urgent and long overdue ethical and socio-sanitary need [32]. From this perspective, the UHC represents a major political milestone towards the SDGs achievement, and a propaedeutic measure to enhance global well-being. Robust public investments in health infrastructures handling preventive, therapeutic and palliative services would ensure the access to the healthcare system to disadvantaged citizens (SDG 1.4), providing a structural response to maternal and neonatal mortality (SDG 3.1 and 3.2); preventing communicable diseases originating in densely populated areas affected by poor hygiene (SDG 3.3); providing answers to people requiring help for therapies against non-communicable diseases (SDG 3.4 and 3.5); and enhancing the standards in low-income countries of sexual and reproductive health-care services (SDG 3.7). Hence, rather than giving a final solution that would immediately improve societal well-being, universal health coverage represents an unalienable milestone creating the preconditions to enhance the impact of healthcare services and eradicate inequalities. This change is mainly relevant for LMICs, where more than 8 million people per year lose their life due to inadequate access to healthcare services [33]. In the LMICs the ex-novo establishment of a UHC regime – and the fulfilment of the SDG target 3.8 – entails higher costs in comparison with the developed countries. In addition to the costs related to the introduction of an array of new health infrastructures and services, the governments of the LMICs require allocation of additional resources to empower human resources, and at the same time, enhancing the capability of higher education institutions to shape highly qualified professionals [34]. Regarding this aspect, improving the national higher education would also mitigate the sanitary and economic damages resulting from the systematic over-prescription and self-prescription of antibiotics in LMICs [35, 36]. Despite the absence of a formal distinction between developed and developing countries in the formulation of the SDGs targets, the establishment of UHC schemes seems mainly addressed to LMICs. It is yet unclear if the economic impact of AMR – and, in some ways, the exclusion of antibiotic resistance from SDG targets – will jeopardise the efforts in the low-income countries to establish a UHC scheme capable of improving well-being and promoting a fairer socio-economic development.

5.4 Antimicrobial resistance and the achievement of Sustainable Development Goals in developing countries

As introduced in the earlier sections, the SDGs and economic policies, programmes, resources are inextricably linked to each other. This implies that the advancement in one area fosters progress in another, and vice-versa a drawback in one of the SDG surely affects (and may impede) the achievement in another. As a result, research investigating the socioeconomic impact of AMR on the ability of the developing countries to attain the SDGs must consider several targets and indicators related to multiple domains.

To assess the consequences of the AMR and ABR in developing countries, the analysis has to begin with the sustainable development goal that is directly related to healthcare services, and more specifically with the targets that more likely will affect the success of the UN 2030 Agenda in the health domain. In LMICs, the establishment of UHC schemes (SDG 3.8) is meant to develop accessible infrastructures providing preventive services, along with extensive therapeutic interventions for communicable and non-communicable diseases. Considering that the frequency of drug-resistant diseases is higher in developing countries [37] – and in the light of the World Bank (2017) estimation portraying a deeper economic loss for LMICs – it is impossible to draft a positive outlook [14]. In the upcoming decades, national and international authorities governing the public health of developing countries will find themselves trapped between the commitment of investing more resources in the public healthcare sector, the incremental rise of the financial resources required to establish a UHC regime, and a declining GDP that will further constraints the budget dedicated to health policies.

Considering the pivotal role played by the UHC in the achievement of the health-related SDGs targets, in developing countries the aggregated socio-economic and sanitary impact of AMR seems a major obstacle for the improvement of societal well-being. In other words, the detrimental impact of the AMR on the public financial capabilities is likely to generate a domino effect jeopardising the achievement of the UN sustainable targets and obstructing the improvement of the healthcare systems in developing countries. For instance, in most African countries maternal and neonatal mortality are declining due to the improved availability of healthcare services. The establishment of health facilities are progressively replacing archaic birth practices, and at the same time, disseminating preventive medical notions that had improved family planning know-how in the community [38]. Similar reasons are also behind the decline of HIV incidence in the African continent. The governments of Rwanda and Ethiopia operated comprehensive investments in the public healthcare sector that have allowed the countries to invert the statistical figures related to the treatment and prevention of communicable diseases [39]. In these countries, it is evident that the financial impact of AMR and ABR could hamper the further development of health facilities, endangering the occurred progress.

Moreover, it must be taken into account that the existence of positive case studies – and overall the improvements observed during the last decades – represents only one side of the coin. Analysing the flip side, health services in developing countries are still a privilege accessible only to a portion of the population. The statistical figures from LMICs denote a structural deficit in the quality of the care services. Only half of the cases of tuberculosis are treated, less than 10 percent of persons suffering from mental health disorders receive support, and the medical assistance to women and children is still far decent [33]. In developing countries, the cases of resistant infective diseases are on the rise, particularly in Sub-Saharan Africa where 60 percent of the HIV-positive patients developed a resistance to the pharmacological therapy [40]; or in South and South-East Asia where antimalarial drugs are gradually losing efficacy [41]. Also in Ethiopia, despite the investments in the public healthcare sector, resistant pneumococcus infections result in the death of nearly 3000 children per annum [42]. Poverty and inequalities remain the minimum common denominator of developing countries. According to UNICEF (2016), the maternal mortality ratio in Sierra Leone in 2015 has been as high as 1 360 deaths per 100 000 births. Moreover, the absence of updated data denotes the inability of the national healthcare system to assess the state of art, and as a logical consequence, to collect and share basic data on the health of the Sierra Leonean population.

The impact of AMR in these communities – besides deteriorating the epidemiological state of the art – will obstruct the establishment of a UHC scheme. Mostly in the least developed countries, public health infrastructures and services represent the only solution to avoid the gradual horizontal corrosion of the community health. In the low-income countries, the impact of the AMR will jeopardise the commitments of the national and international political elites, slowing down the reduction of maternal and neonatal mortality (SDG 3.1-2), the diminution of the global mortality rate for communicable and non-communicable diseases (SDG 3.3-6) – as well as it will exclude millions of women from reproductive and sexual healthcare services (SDG 3.9).

The detrimental impact of AMR and ABR does not end in the health domain. The intersectional analysis of the SDG and AMR reveals several deeper powerful relations that are hidden and may seem irrelevant at a first glance (Figure 5.2.).

In developing countries, the AMR microeconomic impact over citizen income – combined with the macroeconomic consequences of the state financial capabilities to establish a fully-fledged UHC regime – is expected to curb the effectiveness of programs targeting poverty reduction, inequalities, and malnutrition (SDG 1, 2 and 10).

Antimicrobial and antibiotic resistance will push an increasing share of the population into poverty. This will occur for the simple fact that in developing countries individuals managing their household based on a daily wage will immediately enter a status of financial distress once unable to work due to a drug-resistant infection [14]. In contrast with the targets listed under the first, second and tenth SDGs, in the next decade, 18.7 million people in low-income countries will face extreme poverty due to the direct and indirect effects of antimicrobial resistance [43]. Considering the scenario, the disproportioned socio-economic impact of AMR

TABLE 5.2
Expected antimicrobial resistance indirect impact over non-health SDGs
Source: [4]

Elevate detrimental impact		Moderate detrimental impact	
SDG 1	No poverty	SDG 5	Gender equality
SDG 2	Zero hunger	SDG 10	Reduced inequalities
SDG 8	Decent work and economic growth	SDG 12	Responsible consumption and production
SDG 11	Sustainable cities and communities	SDG 14	Life below water
SDG 17	Partnerships for the goals	SDG 15	Life on land

and ABR on low-income counties is expected to generate shocks the labour productivity, increase the unsustainability of the cost of healthcare services both for individuals and the national authorities.

The increased poverty will automatically weaken the efficacy of initiatives tackling global hunger (SDG 2). Malnutrition in developing countries represents one of the major aggravating risk factors for communities that are largely in contact with drug-resistance pathogens [44]. Currently, 811 million people – equivalent to 10% of the world population – do not have regular access to food, and after decades of improvements, during the last two years, global hunger is on the rise [38]. The rise of the world population – along with the economic boom occurring in emerging economies, and the globalisation of consumption patterns – are affecting traditional diets. To meet the increased demand for animal protein and agricultural products, in the upcoming decade food production is projected to grow between 50 and 80 percent [45]. In this context, resistant infections threaten food security and farmers' profits, and simultaneously, consumers fallen under the poverty line would have limited financial resources to purchase groceries. In a nutshell, AMR and ABR are likely to negatively impact both the agricultural productivity and the individual income level, creating concrete obstacles for the successful implementation of the SDG 2 zero hunger.

A similar conclusion can be drawn for the SDG 8 decent work and economic growth and the SDG 11 sustainable cities and communities. In developing countries, the expected detrimental macro and microeconomic effects caused by drug-resistant pathogens represent an obstacle for GDP growth and employment. In parallel, AMR and ABR will horizontally hamper the economic productivity and the sustainability of the developmental achievements. The growing incidence of AMR is also likely to deteriorate the sanitary and economic conditions of people working in the animal farming sector. As highlighted by several studies [46, 47, 48], the spread of drug-resistant diseases among animals is favouring zoonotic processes deteriorating in the first place the health and the financial status of workers employed in intensive animal farms. Moving from the rural to the urban context, socio-economic disadvantaged people in slums – bearing living conditions characterised by poor hygiene, overpopulation, and proximity with livestock – have higher chances to be infected by drug-resistant pathogens [49]. Even in this scenario, people from disadvantaged backgrounds are expected to become the first depository of the negative health and socio-economic consequences of antimicrobial resistance. Mostly if we consider cities in the latest developed countries, this is likely to further hamper the improvement of the housing conditions and the development of sustainable urbanization (SDG 11).

Antimicrobial resistance is likely to have also a moderate hampering impact on the success of the SDG 5 gender equality. In the first place, the financial constrain caused by the AMR and ABR in establishing UHC regimes in low-income will deprive women of reproductive and sexual healthcare services. Moreover, the increasing incidence of resistant pathogens will raise the frequency of infections during pregnancy, abortion, and childbirth [50]. In this scenario, characterised by the absence of a UHC scheme ensuring a minimum standard of hygiene, decent health infrastructures and availability of health professionals, the well-being of women living in low-income countries will be systematically penalised. A further hampering cause is related to the projected rise of individuals requiring care due to prolonged drug-resistant diseases. Considering the absence of affordable healthcare infrastructure and the role played by women within the traditional family scheme [51], the rise of antimicrobial resistance will coerce a higher number of women to take care of family members suffering the consequences of drug-resistant infections, hence reducing at the grassroots level the socio-economic chances for gender empowerment [14].

Linked with the data reported earlier, the targets listed under the responsible consumption and production, life below water and on land goals are likely to be affected by antimicrobial resistance. The increased demand for food supply and the globalisation of regional diets are expected to increase the demand for animal protein. The resulting rising volume of livestock will increase the production and consumption of antibiotics in LMICs [52]. The higher volume of animal and fish intensive farming will negatively impact several indicators measuring production sustainability, creating new challenges in terms of food security for all the actors involved in the production and consumption chain, and at the same time, deteriorating animal health. In addition, as highlighted in several investigations targeting LMICs (e.g.[53, 54, 55]), the enlarged production of antibiotics will further aggravate the pharmaceutical contamination of the water and soil.

Confirming the scale and the horizontal nature of the challenges originating from drug-resistant infections, the World Bank (2019) has included the SDG 17 partnership for the goals among the 2030 Agenda targets that most likely will face severe disruption due to the exponential socio-economic impact of the antimicrobial resistance. The economic consequences of resistant pathogens are expected to generate exorbitant costs also for developed countries. As we saw earlier, the impact of AMR in the US and Europe will exert financial pressure on the pre-existent UHC scheme, and for the public budget tout court. At the same time, low-income countries will be trapped between the commitment to create a UHC regime and the exponential growth of the investment required to respond to the population health needs. In a scenario where developed countries have to face budget constrain and developing countries become even more dependent on international aid, there is the concrete risk to observe a weakening of the investments dedicated to sustainable development.

5.5 Conclusion

The uncontrollable spread of the SARS-CoV-2 epidemic proved to the world that in the realm of health, the principle of leaving no one behind should not be considered a political motto, but an unalienable need for global development policies. Jeopardising the implementation of the UN 2030 Agenda, the cumulative impact of antimicrobial and antibiotic resistance generates a vicious circle composed of interconnected health and socio-economic dynamics. In developing countries, the horizontal impact of AMR on most of the SDGs – besides hampering the establishment of UHC regimes capable to improving the accessibility to health services to people most in need – could create a financial short circuit constraining the impact of the international cooperation programs. From this perspective, the exclusion of AMR and ABR from the SDGs 3 targets appear as an avoidable mistake. This exclusion marginalises a global challenge that, if not addressed, might return the world to a time when public health was threatened by minor infections.

Despite the existence of multiple similarities between antimicrobial resistance and climate change, these key contemporary challenges, rotating around the relation of humankind with nature and the environment, differ on two peculiar aspects. In the first place, while climate change acquired a global resonance, the attention of the media and a representation in politics and civil society, AMR and ABR remain a silent pandemic, largely unknown to most of the population. The second difference represents a ray of hope, and at the same time, it is a call for action. While the solution to climate change implies vast investments and a long-term plan, tackling antimicrobial resistance is relatively cost-effective and achievable within a short time. According to OECD (2018), deaths caused by the AMR could be decreased by 75 percent with an average cost of 2 USD per person a year. Hence, to tackle the devastating impact of drug-resistant pathogens in low-income countries, national governments, and international organisations need to urgently design and implement a global and comprehensive action plan incrementing the financial and intellectual investments. Along with this essential measure, global political elites should also reconsider the role of the state in the pharmaceutical sector, considering the unsustainable implications of leaving the development of new drugs in the hands of private actors and financial considerations.

Bibliography

[1] Poojan Shrestha, Ben S Cooper, Joanna Coast, Raymond Oppong, Nga Do Thi Thuy, Tuangrat Phodha, Olivier Celhay, Philippe J Guerin, Heiman Wertheim, and Yoel Lubell. Enumerating the economic cost of antimicrobial resistance per antibiotic consumed to inform the evaluation of interventions affecting their use. *Antimicrobial Resistance & Infection Control*, 7(1):1–9, 2018.

[2] Oran R Young, Arild Underdal, Norichika Kanie, and Rakhyun E Kim. Goal setting in the anthropocene: The ultimate challenge of planetary stewardship. *Governing through goals: Sustainable Development Goals as governance innovation*, pages 53–74, 2017.

[3] Jasper Littmann, AM Viens, and Diego S Silva. The super-wicked problem of antimicrobial resistance. *Ethics and drug resistance: Collective responsibility for global public health*, 2020.

[4] W Bank. Pulling together to beat superbugs. knowledge and implementaiton gaps in addressing antimicrobial resistance. *World Bank*, 2019.

[5] Jirka Taylor, Marco Hafner, Erez Yerushalmi, Richard Smith, Jacopo Bellasio, Raffaele Vardavas, Teresa Bienkowska-Gibbs, and Jennifer Rubin. Estimating the economic costs of antimicrobial resistance. *Model and Results (RAND Corporation, Cambridge, UK)*, 2014.

[6] Michele Cecchini, Julia Langer, and Luke Slawomirski. Antimicrobial resistance in g7 countries and beyond: economic issues, policies and options for action. *Paris: Organization for Economic Co-operation and Development*, pages 1–75, 2015.

[7] United States. NATIONAL ACTION ANTIBIOTIC-RESISTANT BACTERIA Progress Report : Year 3. (October), 2018.

[8] Thomas P Van Boeckel, João Pires, Reshma Silvester, Cheng Zhao, Julia Song, Nicola G Criscuolo, Marius Gilbert, Sebastian Bonhoeffer, and Ramanan Laxminarayan. Global trends in antimicrobial resistance in animals in low-and middle-income countries. *Science*, 365(6459):eaaw1944, 2019.

[9] Karen L Tang, Niamh P Caffrey, Diego B Nóbrega, Susan C Cork, Paul E Ronksley, Herman W Barkema, Alicia J Polachek, Heather Ganshorn, Nishan Sharma, James D Kellner, et al. Restricting the use of antibiotics in food-producing animals and its associations with antibiotic resistance in food-producing animals and human beings: a systematic review and meta-analysis. *The Lancet Planetary Health*, 1(8):e316–e327, 2017.

[10] Gabriel K Innes, Pranay R Randad, Anton Korinek, Meghan F Davis, Lance B Price, Anthony D So, and Christopher D Heaney. External societal costs of antimicrobial resistance in humans attributable to antimicrobial use in livestock. *Annual review of public health*, 41:141–157, 2020.

[11] Mark Jit, Dorothy Hui Lin Ng, Nantasit Luangasanatip, Frank Sandmann, Katherine E Atkins, Julie V Robotham, and Koen B Pouwels. Quantifying the economic cost of antibiotic resistance and the impact of related interventions: rapid methodological review, conceptual framework and recommendations for future studies. *BMC medicine*, 18(1):1–14, 2020.

[12] Rebecca Knowles, Mike Sharland, Yingfen Hsia, Nicola Magrini, Lorenzo Moja, Amani Siyam, and Elizabeth Tayler. Measuring antibiotic availability and use in 20 low-and middle-income countries. *Bulletin of the World Health Organization*, 98(3):177, 2020.

[13] World Health Organization et al. Who advisory group on integrated surveillance of antimicrobial resistance (agisar): report of the 7th meeting, 17-20 october 2016, raleigh, united

states of america. Technical report, World Health Organization, 2018.

[14] World Bank. *Drug-resistant infections: a threat to our economic future.* World Bank, 2017.

[15] ICMR. Antimicrobial Stewardship Program Guideline, 2018.

[16] Mohammad Ahmad and Asad U Khan. Global economic impact of antibiotic resistance: a review. *Journal of global antimicrobial resistance*, 19:313–316, 2019.

[17] Carolina dos S Ribeiro, Linda HM van de Burgwal, and Barbara J Regeer. Overcoming challenges for designing and implementing the one health approach: A systematic review of the literature. *One Health*, 7:100085, 2019.

[18] Maarten Hajer, Måns Nilsson, Kate Raworth, Peter Bakker, Frans Berkhout, Yvo De Boer, Johan Rockström, Kathrin Ludwig, and Marcel Kok. Beyond cockpit-ism: Four insights to enhance the transformative potential of the sustainable development goals. *Sustainability*, 7(2):1651–1660, 2015.

[19] Mark Woolhouse, Melissa Ward, Bram Van Bunnik, and Jeremy Farrar. Antimicrobial resistance in humans, livestock and the wider environment. *Philosophical Transactions of the Royal Society B: Biological Sciences*, 370(1670):20140083, 2015.

[20] Casey Stevens and Norichika Kanie. The transformative potential of the sustainable development goals (sdgs), 2016.

[21] Michael Anderson, Charles Clift, Kai Schulze, Anna Sagan, Saskia Nahrgang, E Mossialos, et al. Averting the amr crisis: What are the avenues for policy action for countries in europe? 2019.

[22] Elaine Byrne, Brian Nicholson, and Fadi Salem. Information communication technologies and the millennium development goals. *Information Technology for Development*, 17(1):1–3, 2011.

[23] Frank Biermann, Norichika Kanie, and Rakhyun E Kim. Global governance by goal-setting: the novel approach of the un sustainable development goals. *Current Opinion in Environmental Sustainability*, 26:26–31, 2017.

[24] European Antimicrobial Resistance Surveillance System (EARSS et al. Summary of latest data on antibiotic resistance in the european union. *http://www. ecdc. europa. eu/en/activities/surveillance/EARS-Net/Pages/index. aspx*, 2010.

[25] John W McArthur and Krista Rasmussen. Change of pace: Accelerations and advances during the millennium development goal era. *World Development*, 105:132–143, 2018.

[26] Jeffrey D Sachs, Guido Schmidt-Traub, Mariana Mazzucato, Dirk Messner, Nebojsa Nakicenovic, and Johan Rockström. Six transformations to achieve the sustainable development goals. *Nature sustainability*, 2(9):805–814, 2019.

[27] Marlon E Cerf. The sustainable development goals: contextualizing africa's economic and health landscape. *Global Challenges*, 2(8):1800014, 2018.

[28] Michael Marmot and Ruth Bell. The sustainable development goals and health equity. *Epidemiology*, 29(1):5–7, 2018.

[29] Jennifer Prah Ruger and Hak-Ju Kim. Global health inequalities: an international comparison. *Journal of epidemiology & community health*, 60(11):928–936, 2006.

[30] FAOOIE WHO. The fao-oie-who collaboration sharing responsibilities and coordinating global activities to address health risks at the animal-human-ecosystems interfaces. In *International Ministerial Conference on Animal and Pandemic Influenza. Hanoi*, 2010.

[31] Ariel Pablos-Mendez, Karen Cavanaugh, and Caroline Ly. The new era of health goals: universal health coverage as a pathway to the sustainable development goals. *Health Systems & Reform*, 2(1):15–17, 2016.

[32] Harald Schmidt, Lawrence O Gostin, and Ezekiel J Emanuel. Public health, universal health coverage, and sustainable development goals: can they coexist? *The Lancet*, 386(9996):928–930, 2015.

[33] Margaret E Kruk, Anna D Gage, Catherine Arsenault, Keely Jordan, Hannah H Leslie, Sanam Roder-DeWan, Olusoji Adeyi, Pierre Barker, Bernadette Daelmans, Svetlana V Doubova, et al. High-quality health systems in the sustainable development goals era: time for a revolution. *The Lancet global health*, 6(11):e1196–e1252, 2018.

[34] James Campbell, James Buchan, Giorgio Cometto, Benedict David, Gilles Dussault, Helga Fogstad, Inês Fronteira, Rafael Lozano, Frank Nyonator, Ariel Pablos-Méndez, et al. Human resources for health and universal health coverage: fostering equity and effective coverage. *Bulletin of the World Health Organization*, 91:853–863, 2013.

[35] Ann Versporten, Peter Zarb, Isabelle Caniaux, Marie-Françoise Gros, Nico Drapier, Mark Miller, Vincent Jarlier, Dilip Nathwani, Herman Goossens, Andi Koraqi, et al. Antimicrobial consumption and resistance in adult hospital inpatients in 53 countries: results of an internet-based global point prevalence survey. *The Lancet Global Health*, 6(6):e619–e629, 2018.

[36] John W Peabody, Lisa DeMaria, Owen Smith, Angela Hoth, Edmond Dragoti, and Jeff Luck. Large-scale evaluation of quality of care in 6 countries of eastern europe and central asia using clinical performance and value vignettes. *Global Health: Science and Practice*, 5(3):412–429, 2017.

[37] James A. Ayukekbong, Michel Ntemgwa, and Andrew N. Atabe. The threat of antimicrobial resistance in developing countries: Causes and control strategies. *Antimicrobial Resistance and Infection Control*, 6(1):1–8, may 2017.

[38] FAO. Antimicrobial resistance in food and agriculture. Report, 2017.

[39] Paul E Farmer, Cameron T Nutt, Claire M Wagner, Claude Sekabaraga, Tej Nuthulaganti, Jonathan L Weigel, Didi Bertrand Farmer, Antoinette Habinshuti, Soline Dusabeyesu Mugeni, Jean-Claude Karasi, et al. Reduced premature mortality in rwanda: lessons from success. *Bmj*, 346, 2013.

[40] Josep M Llibre. Time to get serious with hiv-1 resistance in sub-saharan africa. *The Lancet Infectious Diseases*, 17(3):241–243, 2017.

[41] Aung Pyae Phyo and François Nosten. The artemisinin resistance in southeast asia: An imminent global threat to malaria elimination. *Towards Malaria Elimination-A Leap Forward*, 2018.

[42] Hui-Han Chen, Andrew Stringer, Tadesse Eguale, Gauri G Rao, and Sachiko Ozawa. Impact of antibiotic resistance on treatment of pneumococcal disease in ethiopia: an agent-based modeling simulation. *The American journal of tropical medicine and hygiene*, 101(5):1042, 2019.

[43] Patrick A Baron, David C Love, and Keeve E Nachman. Pharmaceuticals and personal care products in chicken meat and other food animal products: a market-basket pilot study. *Science of the total environment*, 490:296–300, 2014.

[44] DK Byarugaba. Antimicrobial resistance in developing countries and responsible risk factors. *International journal of antimicrobial agents*, 24(2):105–110, 2004.

[45] Bonnie M Marshall and Stuart B Levy. Food animals and antimicrobials: impacts on human health. *Clinical microbiology reviews*, 24(4):718–733, 2011.

[46] Gijs Klous, Anke Huss, Dick JJ Heederik, and Roel A Coutinho. Human–livestock contacts and their relationship to transmission of zoonotic pathogens, a systematic review of literature. *One Health*, 2:65–76, 2016.

[47] CW Knetsch, N Kumar, SC Forster, TR Connor, HP Browne, C Harmanus, IM Sanders, SR Harris, L Turner, T Morris, et al. Zoonotic transfer of clostridium difficile harboring antimicrobial resistance between farm animals and humans. *Journal of clinical microbiology*, 56(3):e01384–17, 2018.

[48] Nguyen Vinh Trung, Sébastien Matamoros, Juan J Carrique-Mas, Nguyen Huu Nghia, Nguyen Thi Nhung, Tran Thi Bich Chieu, Ho Huynh Mai, Willemien van Rooijen, James Campbell, Jaap A Wagenaar, et al. Zoonotic transmission of mcr-1 colistin resistance gene from small-scale poultry farms, vietnam. *Emerging infectious diseases*, 23(3):529, 2017.

[49] Maya L Nadimpalli, Sara J Marks, Maria Camila Montealegre, Robert H Gilman, Monica J Pajuelo, Mayuko Saito, Pablo Tsukayama, Sammy M Njenga, John Kiiru, Jenna Swarthout, et al. Urban informal settlements as hotspots of antimicrobial resistance and the need to curb environmental transmission. *Nature microbiology*, 5(6):787–795, 2020.

[50] World Health Organization. Countries step up to tackle antimicrobial resistance.

[51] Nidhi Sharma, Subho Chakrabarti, and Sandeep Grover. Gender differences in caregiving among family-caregivers of people with mental illnesses. *World journal of psychiatry*, 6(1):7, 2016.

[52] Bram Van den Bergh, Maarten Fauvart, and Jan Michiels. Formation, physiology, ecology, evolution and clinical importance of bacterial persisters. *FEMS Microbiology Reviews*, 41(3):219–251, 2017.

[53] Mohan Amarasiri, Daisuke Sano, and Satoru Suzuki. Understanding human health risks caused by antibiotic resistant bacteria (arb) and antibiotic resistance genes (arg) in water environments: Current knowledge and questions to be answered. *Critical Reviews in Environmental Science and Technology*, 50(19):2016–2059, 2020.

[54] Dai-Ling Wu, Min Zhang, Lu-Xi He, Hai-Yan Zou, You-Sheng Liu, Bei-Bei Li, Yuan-Yuan Yang, Chongxuan Liu, Liang-Ying He, and Guang-Guo Ying. Contamination profile of antibiotic resistance genes in ground water in comparison with surface water. *Science of the Total Environment*, 715:136975, 2020.

[55] Fangkai Zhao, Lei Yang, Liding Chen, Shoujuan Li, and Long Sun. Co-contamination of antibiotics and metals in peri-urban agricultural soils and source identification. *Environmental Science and Pollution Research*, 25(34):34063–34075, 2018.

6

Disease Economics: Economic aspects of Anti-Bacterial and Anti-Microbial Resistance: Comparative analysis of Europe and other world regions

Romans Putans
Department of International Business and Economics, Riga Stradins University, Latvia

Lauma Springe
Department of Public Health and Epidemiology, Riga Stradins University, Latvia

CONTENTS

6.1 Introduction: Health and Disease Economics

Health care and to that matter even more – a disease care, is completely integral part of modern society, its governance, economics. It is connected to and affects all the spheres of people's lives, all the spheres of economic, and social activities, and thus the ultimate aim – better life, welfare, economic and social security. Economics being organic biological system is driven and kept alive much like a human body – by circulation. Interesting fact showing that we are subjected to and aware of economic system as organic as is the human body, is that the importance of modern economic circulation was discovered comparatively shortly after the discovery of circulation of blood in human body in early 17th century. In fact, in parallel to the development of economic thoughts deriving from mechanical physics – economic "gravity" force pulling the market to the equilibrium, economic "inertia" force representing maximization of profits of economic subjects (people, firms) – the physiocrats pursued the development of economic thought based on medical and health discoveries – they used human body's blood circulation system as an analogue to economic circulation, in which the circulating "economic blood" is essentially goods and services supported by circulation of money and people. In such circulation every asset counts – a broken window is as good for the economy as is e.g., production of milk, because it secures the circulation – provides jobs to windows' producers, woodworkers, carpenters, window repair craftsmen, logistic specialists, who in their turn need other support services, raw materials, equipment, tools, etc. Their clients and customers, in turn, pay them the money that they have earned in exchange for their work with other economic subjects. That is the circulation of many, complex and strongly related resources. Identically, as bad as it sounds psychologically human, also the health and even more a disease is ensuring immensely strong economic circulation – imagine all the variety of medical staff jobs, workplaces, education for them, laboratories, universities, hospital builders, and the very buildings with all what is inside, around, above and beneath them, the development, testing, production, distribution of medicine and health products –

health and disease economics cover circulation of practically all resources in practically all fields of life you can imagine. So, now that we‘ve established the importance role of health and disease in the economics, one might think that it is all right as it is – everything works and fits in its right place and is needed as it is, and if for some reason one or some elements of the system are removed, e.g, the need for health care because people would not get sick, the whole social and economic system would collapse, so everything is needed as it is now. Such a thought, although not entirely true, economically scientific would be close to reasonably justified. The major reason for the reasonability would be because this thought is actually pointing towards the balance, the risks of economic disruption, and the market equilibrium. In the current setting – things being as they are – the market ‘has found‘ its equilibrium point and people are settled in it, used to it. The balance, the equilibrium are the major aspects in the circulation, because it is the gravitational center towards which and around which the circulation takes place, it is the gravitational center that keeps objects steady, in this case – the economic system. But, although we showed how a broken window or an ill person seemed actually needed for the whole economic system, and if there were no broken windows or ill persons, the system would collapse, the equilibrium balance would also collapse if all (or too many) windows were broken or too many people would get ill or could not be treated – too many resources would have to be focused on a specific field thus causing losses and slows the circulation in other fields. So, it is the balance and the equilibrium that is of importance – it affects the circulation. The increasing disease causes economic disruption and slow-down in overall circulation. Disease economics is following and signaling of the trend of change. One of the current threats to the equilibrium balance that is explored in this chapter is that of ABR and AMR. Of course, the world without need of health care is beyond fantasy. The public healthcare budgets and costs form very different, but high shares in all countries‘ budgets and costs – by the share of GDP spent on health care in 2019 the top countries are (17%), (13%), Germany (12%), Armenia (11%), Cuba (11%) followed by HIC EU/EEA countries, UK, Canada, Japan all above 10% of GDP spending on national healthcare. The least share of GDP spent on health care are in Monaco, Djibouti, Congo, Brunei, Papua New Guinea, Benin, Bangladesh, Angola, Lao PDR, Gabon, Kazakhstan, Indonesia, Qatar all below 3% of GDP spent on health care. However, comparing the health care expenditure per capita expresses in US$ the countries‘ ranking differs slightly – top countries by spending on health care per capita are USA (10‘921$), Switzerland (9666$), Norway (8007$), Luxembourg and Denmark (slightly above 6000$), Sweden, Germany, Australia, Netherlands, Austria, Canada (from 5671$ to 5048$) and other HIC. The average health expenditure per capita in HIC is 5736 US$, in OECD member countries – 4948 US$, Euro area – 4001 US$, European Union – 3476 US$. In upper middle-income countries – 555 US$; in Central Europe and Baltics – 1094 US$. In the countries with the least health care expenditure per capita this indicator is as low as 20 US$ per person per year in Madagascar, 25 US$ in Eritrea, 27 US$ in Ethiopia, nevertheless still forming the share of 3-4% of national GDP which is a large share (see Figure 6.1). The average health expenditure per capita in low-income countries is 33 US$, in lower-middle income countries – 95 US$. The average by share of GDP spent on health care in 2019 by the World Bank income groups and economic regions is as follows – High- income – 12.53%, Middle income – 5.33%, Upper-middle income – 5.85, Lower-middle income – 3.71%, Low and middle income – 5.33%, Low income – 4.87% [1], in OECD36 countries – 8,8% (from 16.9% in USA to 3.1.% in Indonesia), EU average – 9,92% (from 11,7% in Germany to 5,74% in Romania and 5,73% in Luxembourg) [2].

It is also important to remark that both of these indicators on public healthcare expenditure, particularly in many middle- and low-income countries may and are varying each year showing growth for the spending on health care per capita up to 15% annually; for some countries the indicator remains unchanged over several years. These indicators clearly show that public healthcare is highly important aspect of countries‘ socioeconomics despite other major socio-economic indicators such as income and general wealth.

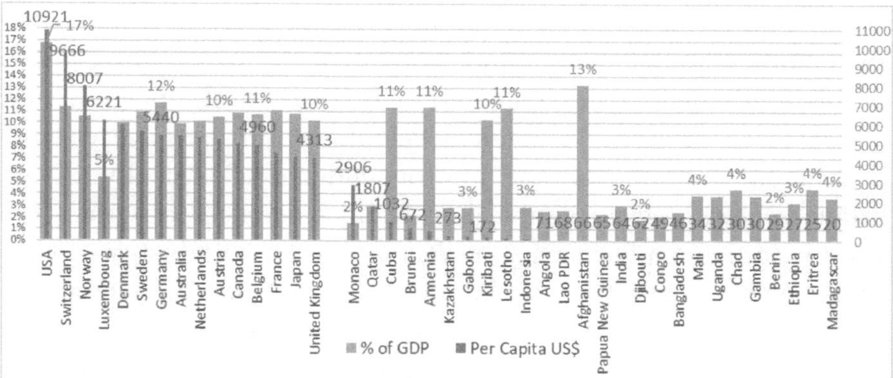

FIGURE 6.1
AMR economics: Health expenditure (in 2019) as percentage of GDP(%) and per capita in US$
Data source: Author's made based on World Bank, World Health Organization, Current health expenditure, 2019 [3]

6.2 ABR and AMR economic impact principles and trends

ABR and AMR economic impact principles and trends are breifly described as follows.

6.2.1 Brief history on building up the economic impact

The discovery of antibiotics and their emerging application in the first half of the, 20th century changed the economic and social landscape of the whole world – the death rates from the respiratory infections decreased substantially in all income levels' countries, people lived longer, worked longer, became more, produced more, specialized more and yes, no need to deny – also wanted and thrived for more. The economic reason why there is absolutely no need to deny the want for more lies in the first part of modern economic theory fundamental problem, which is also physically and organically embedded modern economic paradigm – how to meet the unlimited wants with the scarce resources. The second part relates to the 'invisible hand' which pushes the market to its equilibrium. So, with the anti-biotics more and more people could participate more and more strongly in tackling the economic fundamental problem. Retrospectively we can now, decades later, clearly see that economic development led to the very strong and global interrelation and interdependence of basically all economic sectors – from simple household existence to agriculture, production, manufacturing, healthcare, service industries, waste management, energy, financial markets, information technology, infrastructure, governance, environment and climate – all and each strongly connected and interdependent meaning that changes in one area causes changes in all other areas basically exponentially in geometric progression. In our context it means that despite and paradoxically because of the fact that more and more people could participate more and more strongly in tackling the economic fundamental problem, the very importance, the necessity, and the need of keeping the economic circulation grew exponentially. This need of sustaining the enlarged economic circulation is justified not only by the increased interdependence among various sectors of life, but also by the exponentially increasing number of economic subjects – people and firms, and thus exponentially increasing the already unlimited needs and wants, and the demand of scarce resources. In other words – we are way more people on the planet now – almost 4 times as many as in 1928 when the penicillin was discovered and, as the world population did not grow significantly yet, still almost 4 times as many as in 1940‘s [4] when anti-biotics went in mass usage, and we are becoming more – projected to be almost 5 times as many by and beyond 2050 in comparison to first half of the 20th century [5]. To sustain such a population, we need us ourselves and we need us healthy.

6.2.2 The core of the problems – resistance

As anti-biotics has no one-shot-healthy-for life property, but rather quite the opposite – the more it‘s used the sooner and higher the risks of resistance, thus obviously the major core problem in relation to ABR and AMR impact is, of course, in the R – the resistance. All the rest of the problems and challenges derives from it. The higher levels of drug resistance rates are strongly visibly mostly in low- and middle-income countries [6]. LMIC are subjected to higher drug resistance index (DRI) than HMIC. In LMICs, higher DRI values may reflect a relatively lower level of antibiotic effectiveness due to limited access to newer, more effective antibiotics, revealing where resistance poses a more significant problem [7]. Another equally important aspect is in the comparatively low number of represented countries, including MIC and HIC – only 42 countries are represented; DRI in the abovementioned research is indexed for the countries reporting antibiotic resistance for five or more pathogens and 15 or more pathogen-antibiotic combinations for at least one year between 2012 and 2015 (most recent year) – thus again proving the challenges in data collection, availability and in global national participation in general. Further, three of the major problems, including with economic affect, in relation to the resistance itself are 1) the obvious, that antibiotics envisaged to help encounter with the resistance and are no longer doing their function, 2) the growing trends of discovered resistance and 3) its unpredictability. And these are also the major points where the ABR negative impact and search for solutions starts taking place economically. Although all 3 economically are inseparable, still let us decompose and briefly analyze them separately. First, resistance occurred and discovered, meaning patient is not yet cured as planned (think of millions of patients), patient stays out of active economy, or in case of e.g. a child keeps the caretaker out of the economy causing indirect losses of resources or lost opportunity costs that could be in place (time, labor, money) for creating value in the economy, longer hospital days, longer medical staff hours on the patient, trying of other medicine, increased hospital costs including money, time, labor, increased health care budget costs. Now, imagine hypothetically this happened only once – resistance occurred and discovered – crisis and economic disruption emerged, the alternative health-wise for the patient and in the economy (e.g., substitute worker) is being searched and hopefully found, paid for from e.g. crisis budget and through the crisis turmoil such a hypothetical 'one-time' situation is dealt with, problem solved and the economy with the 'one-time' disruption sooner or later finds its equilibrium again. But second – another resistance discovered. As a result, the shown chain of disruption repeats itself, also meaning the efforts and resources (money, time, labor) medically and economically that are put in search for a solution are doubled, tripled, multiplied as many times as another resistance is discovered. Now, imagine that there are not so may repetitive resistance discovery cases, but somewhat predictable. In addition to the multiplication of resources input we showed for the first resistance discovered or multiplication of hypothetical 'one-time' situation, the repetitive, if not too many, resistances discovered are adjusting the economics in such a way that the repetitive resistance discovered would be foreseen and the resources – money, time, labor – would be planned ahead even if the trend of repetitive resistance discoveries is growing, crucially – to the predictable limit. But here comes the third major problem – as mentioned it is the unpredictability. Furthermore, the current research shows that many estimations of AMR, ABR and its economic impact forecasts should be perceived as underestimation, meaning that the situation is likely to be worth than we anticipate [8].

6.2.3 Problem driver 1 – usage, consumption

The next worrying, disturbing, and immensely impactful factor, which is actually increasing the speed of previously shown problematics, is the increasing usage of antibiotics globally. As antibiotics have clearly shown positive and comparatively quick effect of cure, health, decreased death and illnesses rates, increased life expectancy rates and thus increased actual and potential economic activity and value creation, based on these good examples the usage of antibiotics is increasing globally. Particularly, antibiotic consumption increasing trend is seen in LIC and MIC and their regions, while in HIC, although the very antibiotic consumption is higher in absolute numbers, but the time-dynamic trend show it is even decreasing slightly [9]. The difference in antibiotics consumption trend between LIC and LMIC can be partly explained by the overall quality and efficiency of healthcare systems and thus public health in general, which requires less antibacterial intervention in HIC. But far more important economic scale aspect to look at is in the comparison of a) increasing consumption of antibiotics and b) prevailingly large and increasing population of low- and middle- income countries. LIC and MIC, where the antibiotics consumption trends (the increase of consumption, but not the very consumption yet) are the highest, represent more than 80% of the world's total population. Although the antibiotics are vitally necessary and must be used, the major problem of the increased consumption is that this very consumption is also the major driver of antibiotic resistance. Unfortunately, resistance has eventually been seen to nearly all antibiotics that have been developed [10]. In addition to the resistance discovery itself, it is rather the increasing rates of ABR that bring us back again to previously identified problems and economic distortions like a closed loop. The comparative analyses of antibiotics consumption by countries by income levels shows that in 2015 most of the antibiotics were consumed in high-income countries (top 5 are Spain, Greece, Belgium, France and New Zealand), but the study also showed that consumption in low- and middle-income countries is rapidly converging to rates similar to HICs [6], which, as we'll see later in the chapter, in relation to the increase of global antibiotics, is worrying because low- and middle-income countries represent more than 80% of the world's total population. That means that antibiotics consumption increase can take stronger exponential growth curve. The worrying increasing trend of antibiotic consumption is evident not only in human medical treatment, but also in agriculture – food-producing animals, livestock, fisheries, crops – increased usage being the same problem drivers to the same core problem, the resistance, because the human and veterinary ecosystems are inseparably intertwined and linked [11]. Economic-wise, unlike the antibiotic consumption in human medicine, where the imprudent, inappropriate, and unnecessary usage more often than not may be driven by dubiety, suspense, nescience, carelessness and just in some cases promoted by economic motivators, the antibiotic consumption in animals and farming in addition to its medical function of improving human and animal health by avoiding and curing bacterial infections is often used for purely economic reasons – more efficient production, enhancement of animal growth, unnecessary over-protection [12, 13]. Although it may sound corrupt, still it is very important to note, that the economic drives are also strongly and to large extent justifiably demanded by the economy, by society, people, countries (remember the first part of economics fundamental problem view earlier in this chapter). Nonetheless, the increasing usage and consumption of antibiotics also in the farming is a threat causing risks and disruptions in health and socio-economics.

6.2.4 Problem driver 2 – inappropriate and unnecessary usage and consumption

It is clear that antibiotics are vitally necessary and to an immensely large extent must be used, but the major problem within the increasing consumption of antibiotics is the growing share of the unnecessary or inappropriate antibiotic prescriptions [14]. As shown earlier, detailed global estimates are lacking due to data availability, but for instance in OECD countries according to the WHO report from 2018, as many as half of all antimicrobials used in human health care can be considered inappropriate [15, 16]. Another study carried out by Netherlands institute for health services research under the European Commission financed project estimated that in 2016 7% of the antibiotics taken in the EU were without a prescription (from 2% in Sweden to 20% in Romania and 16% in Greece) [17]. Precise global data collection is problematic as some low- and middle-income countries are under-represented, but research shows that Prescribing practices in low- and middle-income countries are reportedly poor, and robust antibiotic stewardship programmes are often non-existent, e.g., a study of a random sample of hospitals in Vietnam showed that 55% of indications for empirical antibiotic therapy were inappropriate [18]. One of the distinct aspects evident from countries comparison by income level is the indirect circumstances (living conditions, sanitation, hygiene, water, diagnostics) leading to consumption of antibiotics, that would fall under the label of unnecessary, but still some countries are circumstance-forced to use them. For instance, the volume of antibiotic consumption associated with preventable diarrhoeal illness across four middle-income countries (India, Indonesia, Nigeria, and Brazil) is nearly 500 million courses yearly. With universal access to improved water and sanitation it could be reduced by 60% [19]. As a result of imprudent, inappropriate, and unnecessary usage and consumption of antibiotics, not only a comparative waste and opportunity losses of more efficient and needed production in the economy, but it also further facilitates the possibilities and spread of resistance strengthening the described closed loop of economic, social, and medical disruptions (see also Figure 6.2).

6.2.5 Resulting ABR and its economic impact interconnections.

To sum up, as mentioned in previous chapter, it is still a challenge for scientists and for that matter also for countries (practitioners, economists, politians) to quantify ABR and AMR economic costs [20, 21], which we still will try to show further in the chapter, nevertheless the major impact principles and aspects of the ABR and AMR showing clear impact of health issues to economic burden are rather conventionally identified and shared across globe, and they are social and medical intertwined – a) prolonged illness and b) longer hospital days c) additional expensive medicine, d) difficulty to develop new anti-biotics, e) increased death and disabilities [22, 23, 24, 21, 25] causing labor productivity losses and risks national to international economic development investments [26, 27]. All of the mentioned not only disrupts people's ability to participate in the economic circulation around the settled equilibrium point, but also disrupts the very equilibrium and circulation of resources, not to mention the social well-being of these people and their fellow human beings, which has an equally significant impact on induvial and collective economic well-being.

As we saw in previous analysis visualized in scheme above (Figure 6.2) the causes and cores of the ABR problematics, risks, and challenges are evidently not prevailingly solely clinical, medical, and public-healthcare attributed. To an equal, if not larger, extent they are also social (e.g., habits, attitudes, sanitation, and hygiene preconditions and circumstances, education, science, awareness), economic (e.g., equipment, investments, incentives, motivators towards one or another side), political (e.g., regulations, policies, will, awareness, public communication, information management, data availability) and subordinated partly also technological (e.g. to support previously mentioned equipment, information management). Similarly, also the tackling measures are not and cannot

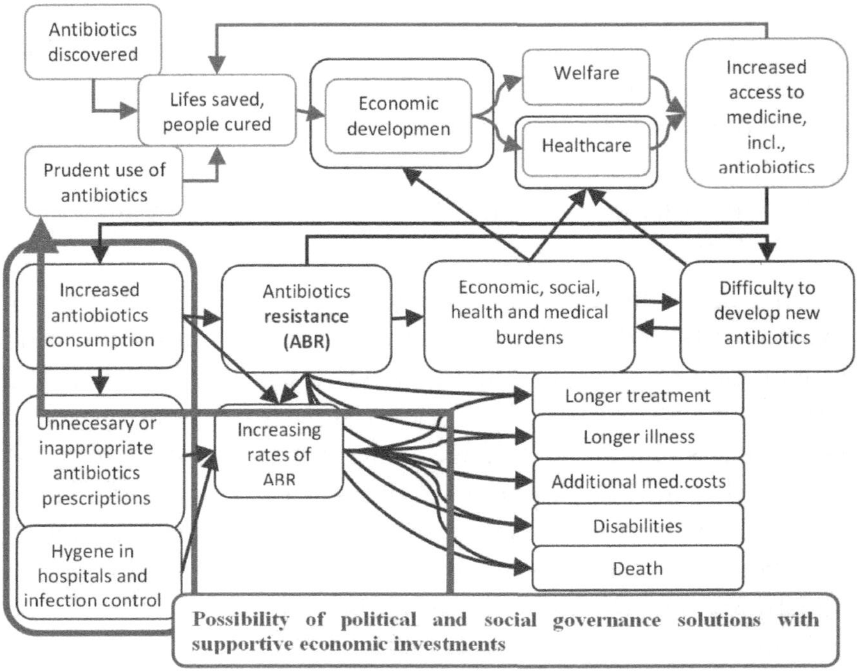

FIGURE 6.2
ABR and economic impact interconnections

be mostly clinical, medical, and public-health-care attributed, but are largely social, economic, and political.

6.3 AMR and ABR economic burden and costs

As mentioned earlier in the book and in this chapter, data availability is a serious challenge in quantifying AMR and ABR burden in all dimensions – medicine, health, social, economic, and thus also political organization, and coordination to tackle the problems which in turn are evident. Parameterizing of AMR and ABR burden, however, is vitally necessary to adjust the performance in, again, all the mentioned dimension to, again – tackle the problems accurately. Thus, although there are AMR and ABR quantified economic burden estimates, which we are going to look at in this part of the chapter, still we'll clearly see that these estimates are of significant variability and further research is essential and ongoing. An important aspect to keep when analyzing the AMR and ABR quantified economic burden estimates is to consider the total AMR caused costs as underestimation, however at some sub-calculation points the AMR and ABR impact quantification is also an overestimation [21]. Another important aspect to keep mind with the quantification of the health and economic burden caused by AMR is that these are to looked at as the additional marginal deaths or costs above the deaths and costs caused by susceptible (i.e, not resistant) to antimicrobial therapies [16] – i.e, the ones that could have been or could be prevented if there were no resistance to antibiotics or in case of projections if the resistance was kept to today's level.

AMR and ABR quantified economic impact estimation derives from the AMR and ABR economic impact principles and indicators we've established earlier in the chapter. One of the most commonly used indicators is productivity loss or reduced labor efficiency, which is expressed in working-age population loss per year, which in turn is estimated mostly by mortality rates and morbidity – the prolonged illness due to ABR, the latter economic-impact-wise is sub-expressed in also commonly used indicator – additional hospital days – allowing the estimation of days and hours not spent in active economy (opportunity costs or loss of opportunity). To the productivity loss or reduced labor efficiency caused by prolonged illness and additional hospital days also the addition health costs are added caused by, e.g, the necessity of treatment of second or third line of antibiotics, which are usually more expensive, as well as economic costs such as, e.g, additional days/hours of medical staff labor to treat the prolonged illness. Further on, the actual available and estimated data on these and other variable indicators are linked to gross domestic product calculations showing the ultimate national, regional, and global economic impact of AMR. In recent years (as of 2016) the estimations of mortality due to drug-resistant infections indicate around 700'000 to as high as 1,29 million people per year [28, 25] and without globally coordinated strong and effective action against AMR might rise more than 10-fold, up to about 10 million AMR caused deaths per year in 2050. Cumulatively this means that 300 million people during the next 35 years are probably to die prematurely because of ABR. Expressed to GDP impact these numbers show as high as 60-100 trillion US$ (2-3,5%) total loss until year 2050. Majority of the losses would take place in low- and middle-income countries in Asia, Africa, and Latin-America, whole, for instance, in OECD countries even in the worst-case scenario of 100% ABR the cumulative loss of OECD countries GDP up to 2050 would be 14,6 trillion US$ [23, 16, 28]. Conversely, delaying the development of widespread antimicrobial resistance by just 10 years could save 65 trillion US$ of the world's output between now and 2050 [19, 29, 30, 23]. For reference – global GDP in 2020 was $\tilde{8}$5 trillion US$ [2], meaning than the economic loss of ABR over 35 years would be somewhat equal to loss of the vitally crucial economic circulation for one year. For another reference – the world's 500 largest companies generated 31.7 trillion US$ in revenues and 1,6 trillion US$ in profits in 2021 (Fortune Global 500). Apart from losses, as for the costs themselves of antimicrobial resistance containment measures as excess to current public health costs, the World Bank estimates it currently at 9 billion US$ per year [31].

6.4 Economic effects of reactive and proactive ABR tackling measures and ideas

The reactive and pro-active measures and proposals to tackle ABR disruptions and further risks can be largely classified in 4 major dimensions – medical-health, economic-financial, social, and policy making (political coordination, organization, regulation, and public governance), the latter also including education (in all dimension) and public communication (mostly, but not limited, in social dimension as the communication in the rest of dimensions would rather be a policy making regulation, but still many public communication benefits may come handy also in medical-health and economic-financial areas addressing the practitioners and investors). The technology would come supportively subordinated in the incentives in major fields are strong, but still, it is important to distinct this subdimension. It is conventionally jointly understood that ABR tackling measures in all of these dimensions cannot be organized separately – it is the crucial point that it takes collective action globally and cross-sectoral [32]. Amongst most often discussed, recommended, implemented, and strengthened ABR tackling measures within wider stewardship and flagship programmes are, e.g., public awareness campaign on national, regional, and global levels, improvement of anti-microbial substance and drug usage and consumption surveillance and data collection, provision, and surveillance globally, pre-screening and more rapid diagnostics, stronger regulations on antibiotic consumption, including that associated with preventable illness to decrease unnecessary use of antibiotics, infection control in hospitals and improved water and sanitation in living areas. As for the latter, research shows that improved water and sanitation would reduce unnecessary use of antibiotics by 60% across middle-income countries in global burden of disease super-regions [28, 16]. A noteworthy aspect and a trend to point out is that large part, although not all, of the ABR tackling measures and challenges are focused on changes to low- and lower-middle income countries, with the high-income countries' role essentially being payments for these global measures. That is justifiably given the a) converging rates and growth of ABR in low- and lower-middle income, b) the obvious income distribution and actual possibilities of financial investments, c) the estimations of economic burden of ABR showing less impact on high- and upper-middle countries. For instance, according to UK government review on antimicrobial resistance [28], a massive global awareness campaign to support programmes mostly in low- and middle-income countries could cost between 40 and 100 million US$ a year in total for high-income countries, which is high cost, of course, but more than reasonable in the prospects of the reviewed risks, losses, and benefits. On the other hand, focusing on financial contribution balance sheet, also indicates a social and political governance separation among regions and countries by income level, which also needs to be looked at prudently as one part of global reforms, where all parties contribute via different means. Also, the awareness campaign is needed in high-income countries, where although the very growth trend of antibiotics consumption and usage is comparatively low and, in some cases, negative (decreasing consumption), but in absolute numbers the consumption, and as shown earlier also the share of imprudent consumption, is higher than in low and lower-middle countries. As for the global costs and economic effects of the tackling measures, as one could guess, they do vary and are scattered across many detailed nuances, sectors, type of drugs, regions, countries, cause-and-effect variables and many other. The UK government commissioned Review on Antimicrobial Resistance provides broad estimate of up to 40 billion US$ over a 10-year period [28] for the cost of global action against AMR – in annual break-down it would take 3-4 billion US$ a year, which is comparatively tiny cost in comparison to previously analyzed worst case scenarios of ABR economic impact of a loss up to 100 trillion US$ (3,3 trillion US$ each ear) until 2050 and costs of hundreds of billion. For the reference of good value for money on a relative basis the 40 billion US$ breakdown in 3-4 billion US$ a year is about 0,05% of what G20 countries spend on healthcare currently (the aggregate G20 healthcare budgets make upto 7 trillion a year) or 20 billion US$ spent in U.S. only in excess costs due to resistance to at first-line antibiotic treatment [28, 33]. Similar estimation is given also by OECD – 2 US$ PPP per capita per year [12] (current OECD population of 1,45 billion x 2 US$ = almost 3 billion US$ or more if 2 US$ calculated by majority of high- and middle-income countries) – such public health package according to OECD could pay for itself within just one year and end up saving USD 4.8 billion US$ per year just in OECD countries themselves. As for the economic incentives for enhanced research and development to new discoveries of antibiotics, the research have shown, they are currently essentially limited to the return of investments intertwined both, medically and economically, meaning that with the growing ABR rates the possibility of new antibiotic discoveries is decreasing clinical-medically; besides the very efforts of new drug discoveries and their trials can even contribute to premature ABR, both of which, of course, decreases of the economic incentives, too [10, 29, 34, 35]. Another health care systematic challenge, medical staff, including pharmaceutical, training, and social responsibility, particularly in low- and middle-income countries, is doctors‘ basic pay, bonus system, which incentivise drug, including anti-biotics, selling including that of ‘over-the-counter‘ [36, 28].

6.5 Conclusion

Health- and disease economics-wise the consequences of AMR issues‘ economic impact principles and factors does not differ much by regions, but does differ further economic effect impact, where high-income countries are significantly better off in taking the economic burden costs caused by AMR implications. AMR economic impact principal indicators expressed in numbers to GDP show as high as 60-100 trillion US$ (2-3,5%) total loss until year 2050. The low- and lower-middle income countries would take majority of losses. Delaying the development of widespread antimicrobial resistance by just 10 years could save 65 trillion US$ of the world's output between now and 2050. Apart from opportunity losses (money not earned), the costs of antimicrobial resistance containment measures as excess to current public health costs, the World Bank estimates currently at 9 billion US$ per year. The UK government commissioned Review on Antimicrobial Resistance provides broad estimate of up to 40 billion US$ over a 10-year period for the cost of global actions tackling AMR risks and disruptions – in annual break-down it would take 3-4 billion US$ a year. Similar estimation is given also by OECD – 2 US$ PPP per capita per year – such public health package according to OECD could pay for itself within just one year and end up saving USD 4.8 billion US$ per year just in OECD countries themselves. Antibiotics consumption and ABR data collection, provision, and surveillance locally and globally is still a challenge thus also transferred to economic estimations. Public health and social dimension wise, although in absolute numbers, currently high-income countries are using more antibiotics than low- and lower-middle countries, they actually need it less (due to better health care system, sanitation, hospital hygiene, infection control), so the share of unnecessary consumed antibiotics might be currently higher in high-income countries. Low- and middle-income countries are rapidly converging to these rates and are likely to surpass them. In high-income countries improper use drives the ABR more strongly than in low- and middle-income countries, where the major resistance driver is also necessary increased access to antibiotics. One of the very strong recommendations in all ABR related research is the need of further and detailed research which the authors of this chapter and this book strongly support.

Bibliography

[1] World Health Organization. Antimicrobial resistance, 2021.

[2] Robin van Kessel, Brian Li Han Wong, Ivan Rubinić, Ella O'Nuallain, and Katarzyna Czabanowska. Is europe prepared to go digital? making the case for developing digital capacity: an exploratory analysis of eurostat survey data. *PLOS Digital Health*, 1(2):e0000013, 2022.

[3] World Health Organization. Antimicrobial Stewardship Programme in Health-Care Facilities in Low and Middle Income Countries, 2019.

[4] World Bank. *World development report 2021: Data for better lives*. The World Bank, 2021.

[5] IACG United. No time to wait–securing the future from drug-resistant infections. *Report to the Secretary General of the Nations*, 2019.

[6] Eili Y Klein, Thomas P Van Boeckel, Elena M Martinez, Suraj Pant, Sumanth Gandra, Simon A Levin, Herman Goossens, and Ramanan Laxminarayan. Global increase and geographic convergence in antibiotic consumption between 2000 and 2015. *Proceedings of the National Academy of Sciences*, 115(15):E3463–E3470, 2018.

[7] World Health Organization et al. Tripartite and unep support ohhlep's definition of "one health". 2021, 2021.

[8] Klaus Kaier and Uwe Frank. Measuring the externality of antibacterial use from promoting antimicrobial resistance. *Pharmacoeconomics*, 28(12):1123–1128, 2010.

[9] Matthew D Brown, Tim K Mackey, Craig N Shapiro, Jimmy Kolker, and Thomas E Novotny. Bridging public health and foreign affairs: the tradecraft of global health diplomacy and the role of health attachés. *Sci Diplomacy*, 3(3):1–12, 2014.

[10] C Lee Ventola. The antibiotic resistance crisis: part 1: causes and threats. *Pharmacy and therapeutics*, 40(4):277, 2015.

[11] Avantika Mann, Kiran Nehra, JS Rana, and Twinkle Dahiya. Antibiotic resistance in agriculture: Perspectives on upcoming strategies to overcome upsurge in resistance. *Current Research in Microbial Sciences*, 2:100030, 2021.

[12] Ramanan Laxminarayan, Thomas Van Boeckel, and Aude Teillant. The economic costs of withdrawing antimicrobial growth promoters from the livestock sector. 2015.

[13] Jonathan Anomaly. Antibiotics and animal agriculture: The need for global collective action. In *Ethics and drug resistance: Collective responsibility for global public health*, pages 297–308. Springer, Cham, 2020.

[14] Barbara Bortone, Charlotte Jackson, Yingfen Hsia, Julia Bielicki, Nicola Magrini, and Mike Sharland. High global consumption of potentially inappropriate fixed dose combination antibiotics: Analysis of data from 75 countries. *PLoS One*, 16(1):e0241899, 2021.

[15] World Health Organization. Countries step up to tackle antimicrobial resistance.

[16] Michele Cecchini, Julia Langer, and Luke Slawomirski. Antimicrobial resistance in g7 countries and beyond: economic issues, policies and options for action. *Paris: Organization for Economic Co-operation and Development*, pages 1–75, 2015.

[17] EA Ashley, J Recht, A Chua, D Dance, M Dhorda, N Thomas, et al. Antimicrobial resistance in low and middle income countries: an analysis of surveillance networks, 2017.

[18] Raph L Hamers and H Rogier van Doorn. Antibiotic consumption in low-income and middle-income countries. *The Lancet Global Health*, 6(7):e732, 2018.

[19] J O'neill. Antimicrobial resistance. *Tackling a crisis for the health and wealth of nations*, 2014.

[20] Susanna J Dunachie, Nicholas PJ Day, and Christiane Dolecek. The challenges of estimating the human global burden of disease of antimicrobial resistant bacteria. *Current Opinion in Microbiology*, 57:95–101, 2020.

[21] Mark Jit, Dorothy Hui Lin Ng, Nantasit Luangasanatip, Frank Sandmann, Katherine E Atkins, Julie V Robotham, and Koen B Pouwels. Quantifying the economic cost of antibiotic resistance and the impact of related interventions: rapid methodological review, conceptual framework and recommendations for future studies. *BMC medicine*, 18(1):1–14, 2020.

[22] Oguz Resat Sipahi. Economics of antibiotic resistance. *Expert review of anti-infective therapy*, 6(4):523–539, 2008.

[23] Jirka Taylor, Marco Hafner, Erez Yerushalmi, Richard Smith, Jacopo Bellasio, Raffaele Vardavas, Teresa Bienkowska-Gibbs, and Jennifer Rubin. Estimating the economic costs of antimicrobial resistance. *Model and Results (RAND Corporation, Cambridge, UK)*, 2014.

[24] Philip Taylor and Robert Reeder. Antibiotic use on crops in low and middle-income countries based on recommendations made by agricultural advisors. *CABI Agriculture and Bioscience*, 1(1):1–14, 2020.

[25] Christopher JL Murray, Kevin Shunji Ikuta, Fablina Sharara, Lucien Swetschinski, Gisela Robles Aguilar, Authia Gray, Chieh Han, Catherine Bisignano, Puja Rao, Eve Wool, et al. Global burden of bacterial antimicrobial resistance in 2019: a systematic analysis. *The Lancet*, 2022.

[26] Ramanan Laxminarayan, Adriano Duse, Chand Wattal,

Anita KM Zaidi, Heiman FL Wertheim, Nithima Sumpradit, Erika Vlieghe, Gabriel Levy Hara, Ian M Gould, Herman Goossens, et al. Antibiotic resistance—the need for global solutions. *The Lancet infectious diseases*, 13(12):1057–1098, 2013.

[27] Katharina Hauck. The economics of infectious diseases. In *Oxford Research Encyclopedia of Economics and Finance*. 2018.

[28] Jim O'Neill. Tackling drug-resistant infections globally: final report and recommendations. 2016.

[29] Aastha Chokshi, Ziad Sifri, David Cennimo, and Helen Horng. Global contributors to antibiotic resistance. *Journal of global infectious diseases*, 11(1):36, 2019.

[30] Porooshat Dadgostar. Antimicrobial resistance: implications and costs. *Infection and drug resistance*, 12:3903, 2019.

[31] W Bank. Pulling together to beat superbugs. knowledge and implementaiton gaps in addressing antimicrobial resistance. *World Bank*, 2019.

[32] Debra A Goff, Ravina Kullar, Ellie JC Goldstein, Mark Gilchrist, Dilip Nathwani, Allen C Cheng, Kelly A Cairns, Kevin Escandón-Vargas, Maria Virginia Villegas, Adrian Brink, et al. A global call from five countries to collaborate in antibiotic stewardship: united we succeed, divided we might fail. *The Lancet Infectious Diseases*, 17(2):e56–e63, 2017.

[33] Richard Smith and Joanna Coast. The true cost of antimicrobial resistance. *Bmj*, 346, 2013.

[34] E Power. Impact of antibiotic restrictions: the pharmaceutical perspective. *Clinical Microbiology and Infection*, 12:25–34, 2006.

[35] Matthew J Renwick, David M Brogan, and Elias Mossialos. A systematic review and critical assessment of incentive strategies for discovery and development of novel antibiotics. *The Journal of antibiotics*, 69(2):73–88, 2016.

[36] Marcella Alsan, Lena Schoemaker, Karen Eggleston, Nagamani Kammili, Prasanthi Kolli, and Jay Bhattacharya. Out-of-pocket health expenditures and antimicrobial resistance in low-income and middle-income countries: an economic analysis. *The Lancet infectious diseases*, 15(10):1203–1210, 2015.

7

Leveraging Health Diplomacy in achieving AMR Policy Coherence

Sanjay Pattanshetty

Department of Health Policy, Prasanna School of Public Health, Manipal Academy of Higher Education, Manipal, Karnataka, 576104, India

Helmut Brand

Department of International Health, Care and Public Health Research Institute – CAPHRI, Maastricht University, Maastricht, The Netherlands

CONTENTS

7.1 Introduction

Antimicrobial resistance (AMR) is an international Public Health issue of concern due to its significant impact on diverse sectors and welfare of communities' worldwide. Rapid increase in AMR burden and spread, and its danger to economic growth and national health security, have brought AMR to the forefront of high-level geo-political discussions in various forums - G7, G20, World Health Assembly and United Nations General Assembly. AMR is now both health and foreign policy priority. Member States during the high-level meeting of the General Assembly at United Nations adopted "Political declaration on antimicrobial resistance" in accordance with United Nations General Assembly resolution. Member states reiterated that AMR will impact the progress made in the prevention and control of infectious diseases. AMR impact on public health will also impact the progress of 2030 Agenda for Sustainable Development. (United Nations 2016). Every year, children die from preventable infections due to lack of access to antimicrobials. Ensuring access is emphasized in target 8 of SDG3: "Achieving universal health coverage including access to safe, effective, quality, and affordable essential medicines for all" [1].

The direct consequences of AMR can lead to increase in mortality and morbidity due to severity of infections, longer hospitalization, increase in cost and decrease in quality of life. AMR is a major threat to human, plant and animal health around the world. The indirect consequences of AMR extend beyond health risks and health policies [2]. Food, Security, Trade, Economy and Environmental sectors are also impacted either directly or indirectly due to the global burden of AMR. AMR is an inter-sectoral issue impacting key global governance sectors and actors. Member states "acknowledge that the resistance of bacterial, viral, parasitic and fungal microorganisms to antimicrobial medicines that were previously effective for treatment of infections is mainly due to the inappropriate use of antimicrobial medicines in the public health, animal, food, agriculture and aquaculture sectors; lack of access to health services, including to diagnostics and laboratory capacity; and antimicrobial residues in soil, crops and water" [3]. As the threat posed by AMR increases the need for policy coherence among different sectors and actors becomes vital.

AMR is a global problem and the solutions have to be implemented collectively. The complicated, multi-sectoral and multi-stakeholder nature of the AMR calls for a strengthened and formal mechanisms to coordinate the global response to AMR. In addition to technical competency to address problems related to AMR, it is important to emphasize the need for transnational multi-stakeholder discussions to find appropriate solutions to address AMR. Global health diplomacy (GHD) is one such approach that can facilitate the involvement of diverse policy makers and sectors at different levels of governance. This chapter argues that there is a need for policy coherence in promoting stewardship, conservation, access and innovation in existing and newer antimicrobials. A successful policy must strike the right balance between three conflicting goals of access, conservation, and innovation.

7.2 AMR Policy Initiatives

World Health Organization (WHO), Food and Agriculture Organization (FAO) and World Organization for Animal Health (OIE) adopted policy initiatives to tackle AMR. WHO/FAO/OIE action plans are mentioned in Table 7.1 and policy avenues in Table 7.2.

7.3 Policy Coherence: Policy framing, Policy Actors and Sectors in AMR

Policy coherence depends largely on policy framing of a problem by different policy actors and sectors. As mentioned in Table 7.1 and 7.2 WHO/FAO/OIE agreed on a plan to address antimicrobial resistance [5]. These organizations are also the standard-setting organizations. However, it is important to note the involvement of

TABLE 7.1
WHO/FAO/OIE action plans
Source: [4]

2015 WHO Global Action Plan on Antimicrobial Resistance	2016 OIE Strategy on Antimicrobial Resistance and the Prudent Use of Antimicrobials
1. Improve awareness and understanding of AMR through effective communication, education and training. 2. Strengthen the knowledge and evidence base through surveillance and research. 3. Reduce the incidence of infection through effective sanitation, hygiene and infection prevention measures. 4. Optimize the use of antimicrobial medicines in human and animal health. 5. Develop the economic case for sustainable investment that takes account of the needs of all countries, and increase investment in new medicines, diagnostic tools, vaccines and other interventions.	1. Improve awareness and understanding. 2. Strengthen knowledge through surveillance and research. 3. Support good governance and capacity building. 4. Encourage implementation of international standards. **2016 FAO Action Plan on Antimicrobial Resistance** 1. Improve awareness of AMR and related threats. 2. Develop capacity for surveillance and monitoring of AMR and AMU in food and agriculture. 3. Strengthen governance related to AMU and AMR in food and agriculture. 4. Promote good practices in food and agricultural systems and the prudent use of antimicrobials

TABLE 7.2
Policy avenues for Averting the AMR crisis
Source: [4]

Policy avenues	Measures
Strengthening surveillance and monitoring, and moving towards national 'One Health' surveillance systems	Strengthen national AMR surveillance system
Strengthening antimicrobial stewardship in human health	To reduce inappropriate prescribing in primary care and hospital care settings (Educational, Persuasive, restrictive and enabling environment
Strengthening infection prevention and control (IPC) in human health	To reduce healthcare-associated infection (HAI) rate (about 30–40% of HAIs are caused by resistant organisms)
Strengthening IPC and reducing inappropriate antibiotic use in animals	Restricting the use of non-therapeutic antimicrobials and antimicrobials that are critically important for human use. IPC measures such as implementing higher biosecurity, improved husbandry methods, use of vaccination.
Limiting the exposure of antimicrobial-resistant pathogens to the environment	Monitoring and regulation of the environment
Fostering R&D of new antimicrobial therapies, diagnostics and vaccines	R&D in antibiotics; R&D in diagnostics: R&D in vaccines; Harmonization of regulatory standards and procedures Coordinating research

other sectors/norms beyond health sector in AMR prevention and control. To mention few sectors that play a major role in ensuring safety and standards of plant and animal products are related to trade agreements such as Sanitary and Phytosanitary(SPS) Agreements, Technical Barriers to Trade (TBT), and International Convention on the Harmonized Commodity Description and Coding System (HCDCS), etc. Access and innovation to antimicrobials is impacted by the norms negotiated in Trade Related Intellectual Property Rights (TRIPS) [6]. Ensuring access to affordable medicines using TRIPS provisions to support research and development of medicine and vaccines is articulated in the Sustainable Development Goal 3. The declaration adopted in United Nations in 2016 "recognise the importance of delinking the cost of investment in research and development on antimicrobial resistance from the price and volume of sales so as to facilitate equitable and affordable access to new medicines, diagnostic tools, vaccines." Harmonization of AMR policies among actors and sectors in health and trade is important to balance the utilization of antimicrobials among human, plant and food animals. Standard-setting organizations such as WHO, FAO and OIE in cooperation with WTO, World Intellectual Property Organization (WIPO) and others can influence policy coherence and regulations to tackle antimicrobial resistance (Figure 7.3.).

7.4 Policy challenges in the prevention and control of AMR from the context of policy frame

Like microorganisms, AMR has no borders. AMR is a borderless threat requiring coordination among actors and sectors to initiate mechanisms to prepare, prevent, alleviate AMR burden globally. The past attempts to prevent AMR has failed for couple of reasons. Uninterrupted increase in global over consumption of antimicrobials and subsequent proliferation of resistance conferring genes. There was 46% increase in antibiotic consumption since 2000. This increase in antibiotic consumption highlights the increasing threat of antibiotic resistance due to misuse of antibiotics among ani-

TABLE 7.3
Policy measures to prevent AMR
Source: [6]

Measures	Trade	Health related
To improve *access* to antimicrobials		
Develop equitable pricing and licensing models to support access to antimicrobials in resource limited settings	TRIPS	WHO, FAO and OIE
End the manufacturing, sale and export of substandard, falsified and banned antimicrobial products	TBT, HCDCS	WHO, FAO and OIE
Strengthen health systems to improve infection prevention and control and reduce the need for antimicrobials	TBT	WHO
On the *conservation* front		
Strengthen surveillance of antimicrobial use and antimicrobial resistance in humans, animals and the environment	SPS	WHO, FAO and OIE
Promote the responsible use of antimicrobials in humans	TBT, SPS, TRIPS	IHR, WHO, FAO, OIE
Promote the responsible use of antimicrobials in animals and agriculture	SPS	WHO, FAO and OIE
Safeguard the effectiveness of newly developed antimicrobials	SPS, TBT, TRIPS	WHO, FAO and OIE
Limit antimicrobial contamination in the environment	SPS	FAO and OIE
For *innovation*		
Incentivize the development of new antimicrobials and related technologies	TRIPS, TBT, HCDCS	IHR
Facilitate the development of new antimicrobials and related technologies	TRIPS	IHR

mals and humans [7]. Nonetheless, there was 39% reduction in antibiotic-resistant bacteria among animals due to interventions that promoted the restriction of antimicrobial use among food producing animals [8]. The Tripartite cooperation between standard setting organizations such as WHO, FAO and OIE has provided stewardship to countries in preventing antimicrobial resistance in recent years but remains seriously under-resourced. The role and support of UN Environment Program (UNEP) to member's states, WHO, FAO and OIE is vital in addressing antimicrobial resistance within environmental policy-making [6]. There is need for concerted and consistent action among different stakeholders. A lack of concerted action due to conflicting interests and regulations is also impacting AMR prevention and control. Both competition and convergence in policy can arise from differential framing of a policy issue. Due to multi-faceted nature of AMR, actors would see AMR as a problem from different policy lens and initiate solutions that are specific to one sector. For instance, focus is mainly on patients and diagnostics if AMR is seen as a healthcare issue. Doctors, and other healthcare professionals have different perspective on origin of AMR as a problem, its significance in the context of patient outcomes and policy initiatives that are needed to halt the progress of AMR. WHO and Ministries of Health are seen as main actors in this context. Similarly, OIE, FAO, Ministry of Agriculture and Health are perceived as main actors responsible to initiate policy activities if AMR is framed as One Health Issue [9]. WTO and WIPO frame AMR issue from the policy lens of innovation, market access, trade and investments. International trade agreements influence AMR and there is a need for understanding of the role and impact of counterfeit trade in the utilization of antimicrobials in animals [10]. Animals tested positive for resistant strains will have direct impact on trade. Availability and accessibility of antimicrobials for human and animal consumption is affected due to trade in antimicrobials [1]. SPS Measures allow countries to set their own/high standards for food safety and animal and plant health which are less trade restrictive, transparent, based on scientific justification, appropriate assessment of risks, international standards, and guidelines. The SPS agreement has provisions for countries to adapt different measures to inspect products due to differences in environmental conditions, prevailing diseases, determinants of diseases in the country of origin with an understanding that the adapted guidelines provide the same level of health protection [11]. However, the challenge is to balance conflicting policies on standards and methods that are trade restrictive between different countries. For instance, Argentina in the year 2018 raised concerns regarding "the European Union's (EU) proposed regulation on veterinary medicinal products, stating that the adoption of provisions regarding the use of antimicrobials in the veterinary sector would have a significant impact on international trade." Argentina was concerned that the proposed text is trade restrictive, lacks transparent and technical basis. The proposed EU regulation would prevent animal products from Argentina to access EU market. AMR is a concern raised in Free Trade Agreements (FTA) and Regional Trade Agreements (RTA). For instance, AMR is a concern in the Transatlantic Trade and Investment Partnership (T-TIP) negotiations. The EU has higher standards for antibiotic use in food animal production. However, 80% of antibiotics sold in the United States of America are used in food animal production. The current variations in standards and regulations were the cause for noticeable concern in FTA and RTA such as Transatlantic Trade and Investment Partnership (T-TIP) negotiations. If not ratified, lowering antibiotic use standards to promote trade would pose increasing risk of AMR in Atlantic region ("Why Transatlantic Trade Must Play a Role in Addressing Antibiotic Resistance" 2014). Similar AMR concerns were expressed by Australia in Trans-Pacific Partnership (Anna George 2016). Mr Ivan McKee (Trade minister) iterated similar concern in the UK-Japan FTA. He emphasized the need for UK government to strengthen the negotiations to tackle AMR. Trade and investment agreements promoted in bilateral and regional FTAs can enable corporations to challenge standards regulating AMR which may compromise AMR strategies. The joint symposium on AMR, held by WTO, WIPO and the WHO in October 2016, highlighted the significance of global cooperation in fostering innovation, access, conservation, and appropriate us of newer antibiotics. From a foreign policy perspective, determinants of AMR are transnational. Supply chain management is complex and major pharmaceutical companies have multi-national presence. It is imperative to initiate collective action, and coordinated international responses to tackle AMR adhering to standards, guidelines and recommendations given by standard setting organizations [12].

7.5 Policy Solution – Leveraging Global Health Diplomacy

AMR is rapidly emerging as a major international public health problem. To mitigate this challenge, a concerted action is needed among different actors and sectors in governments and international organizations. AMR is a health and foreign policy issue in the context of concerns to national and international health security. There is increasing threat due to preventable infections and re-emerging infectious diseases. Emerging and re-emerging diseases will impact costs and benefits of 'securitizing' health [13, 12]. An effective response will require shared transnational action and diplomatic coordination at the global level to meet the challenges of innovation, conservation, and access to antimicrobials. The United Nations declaration on AMR underscored the significance of political engagement at all levels to prevent AMR. Global health diplomacy can play a significant role in ensuring political engagement through negotiation, agreements, and inter-sectoral collaboration between diverse and complex sectors and actors involved in AMR prevention and control. Global Health Diplomacy as an international tool was used among member states and other actors to address global health policy challenges. For instance, The International Health Regulations (IHR) and the Framework Convention on Tobacco Control (FCTC) are examples of international health policy making with certain binding norms that have been most widely debated and researched by different actors and sectors [12]. However, current AMR control measures are mostly non-binding in nature such as - norms, standards, global action plan on AMR and political declaration of the High-Level Meeting on AMR. In addition, AMR prevention efforts are daunted by the complexity in the current global governance system that impact sectors beyond health sector, including the regulations governing international trade, security, development, environment, innovation, and intellectual property. In this context, health diplomacy as a tool would provide a structure and platform for negotiation among partners to promote mutual cooperation and collaboration to prioritise AMR prevention and control.

We have developed a modified Global Health Diplomacy Pyramid highlighting the importance of interaction at different levels that can play a significant role in the prevention and control of AMR [14]. We believe that core health diplomacy measures will have impact on AMR at global level similar to IHR, FCTC and climate negotiations/treaty. Sectoral health diplomacy is important but the impact could be restricted to a sector where intervention is implemented. In addition, it would pose a danger of lack of compliance to norms by individual actors/sectors impacting the effectiveness of global efforts (Figure 7.1.).

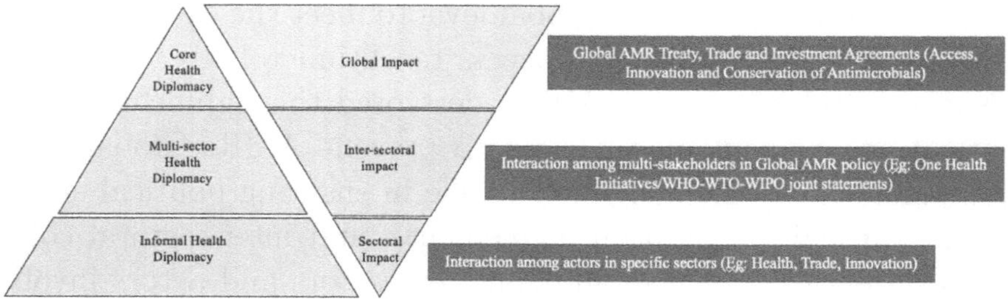

FIGURE 7.1
AMR global health diplomacy pyramid
Source: Author's own elaboration from the literature on Global Health Diplomacy

7.6 Conclusion

In conclusion, collaborative and collective policy action is needed to tackle AMR. Aligning the AMR policies with the SDG goals to ensure access to safe, effective, quality, and affordable essential medicines for all needs inter-sectoral and international commitment. Enabling global health policy environment to implement population level targeted strategies to address AMR is needed. In this context, health diplomacy measures can play a significant role in promoting cooperation and collaboration in harmonization of policy landscape of AMR.

Bibliography

[1] Johanna Hanefeld, Mishal Khan, Göran Tomson, and Richard Smith. Trade is central to achieving the sustainable development goals: a case study of antimicrobial resistance. *BMJ*, 358, 2017.

[2] Christy Manyi-Loh, Sampson Mamphweli, Edson Meyer, and Anthony Okoh. Antibiotic Use in Agriculture and Its Consequential Resistance in Environmental Sources: Potential Public Health Implications. *Molecules : A Journal of Synthetic Chemistry and Natural Product Chemistry*, 23(4), 2018.

[3] UNICEF et al. The state of food security and nutrition in the world 2021. 2021.

[4] Michael Anderson, Charles Clift, Kai Schulze, Anna Sagan, Saskia Nahrgang, E Mossialos, et al. Averting the amr crisis: What are the avenues for policy action for countries in europe? 2019.

[5] World Health Organization. Countries step up to tackle antimicrobial resistance.

[6] Susan Rogers Van Katwyk, Alberto Giubilini, Claas Kirchhelle, Isaac Weldon, Mark Harrison, Angela McLean, Julian Savulescu, and Steven J Hoffman. Exploring models for an international legal agreement on the global antimicrobial commons: lessons from climate agreements. *Health Care Analysis*, pages 1–22, 2020.

[7] Annie J Browne, Michael G Chipeta, Georgina Haines-Woodhouse, Emmanuelle PA Kumaran, Bahar H Kashef Hamadani, Sabra Zaraa, Nathaniel J Henry, Aniruddha Deshpande, Robert C Reiner Jr, Nicholas PJ Day, et al. Global antibiotic consumption and usage in humans, 2000–18: a spatial

modelling study. *The Lancet Planetary Health*, 5(12):e893–e904, 2021.

[8] Karen L Tang, Niamh P Caffrey, Diego B Nóbrega, Susan C Cork, Paul E Ronksley, Herman W Barkema, Alicia J Polachek, Heather Ganshorn, Nishan Sharma, James D Kellner, et al. Restricting the use of antibiotics in food-producing animals and its associations with antibiotic resistance in food-producing animals and human beings: a systematic review and meta-analysis. *The Lancet Planetary Health*, 1(8):e316–e327, 2017.

[9] Didier Wernli, Peter S Jørgensen, Chantal M Morel, Scott Carroll, Stephan Harbarth, Nicolas Levrat, and Didier Pittet. Mapping global policy discourse on antimicrobial resistance. *BMJ Global Health*, 2(2):e000378, 2017.

[10] Tim K Mackey, Bryan A Liang, Peter York, and Thomas Kubic. Counterfeit drug penetration into global legitimate medicine supply chains: a global assessment. *The American journal of tropical medicine and hygiene*, 92(Suppl 6):59, 2015.

[11] World Trade Organization. Understanding the wto agreement on sanitary and phytosanitary measures. 1998.

[12] Brad Spellberg, Gail R Hansen, Avinash Kar, Carmen D Cordova, Lance B Price, and James R Johnson. Antibiotic resistance in humans and animals. *NAM Perspectives*, 2016.

[13] Porooshat Dadgostar. Antimicrobial resistance: implications and costs. *Infection and drug resistance*, 12:3903, 2019.

[14] Matthew D Brown, Tim K Mackey, Craig N Shapiro, Jimmy Kolker, and Thomas E Novotny. Bridging public health and foreign affairs: the tradecraft of global health diplomacy and the role of health attachés. *Sci Diplomacy*, 3(3):1–12, 2014.

8

AMR policies and Implementation issues: Developed vs Developing Countries

CONTENTS

8.1 Introduction

Every country in the world both developed and developing is passing through the epidemiological transition of diseases. This can be described under three major heads: 1. Control of existing infectious diseases, 2. Infectious illnesses that are developing or re-emerging, and 3. Rapid increase in non-communicable diseases (NCDs). In developing countries although the burden of infectious diseases is coming down slowly, yet there are many infectious diseases still prevalent and need to be treated and controlled. In high-income nations, infectious and parasitic illnesses account for 2.3 percent of total DALYs, but they account for 23.1 percent in low- and middle-income countries [1]. In developed countries, most of infectious diseases have been controlled and their main problem is non-communicable diseases. However, the emerging diseases like SARS, Covid-19 is changing the scenario of the infectious disease burden. The advent of antimicrobials has helped both developing and developed countries in treating and controlling infectious diseases. This achievement is now under threat because of development of resistance by the microbes to the antimicrobials [2]. All countries are affected by AMR, however the burden is highly disproportionate in poor and middle-income countries.The AMR epidemiological transition covers 1. Tackling the existing infections resistance to commonly available antimicrobials; 2. Preventing AMR in emerging infections; 3. Sustaining the effectiveness of the existing antimicrobials; and 4. Developing newer antimicrobials specifically antibiotics. The reasons of antimicrobial resistance (AMR) in developing nations are diverse, and may be anchored in health care providers' and patients' attitudes about antimicrobial usage, as well as antimicrobial supply networks in the community. A few of these factors may include inappropriate prescription practises, such as the widespread use of antibiotics during the Covid19 pandemic, insufficient patient education, limited diagnostic facilities, unauthorised antimicrobial sales, a lack of appropriate functioning drug regulatory mechanisms, and non-human use of antimicrobials, such as in animal production [2]. In both developed and developing countries, the human factors of AMR may vary in intensity and magnitude calling for country specific AMR containment action plan. Given the gravity of the AMR crisis, the World Health Assembly endorsed a global action plan on antimicrobial resistance in 2015, directing member countries to develop and implement country-specific AMR action plans [3]. As there are differences in the epidemiological situations in the infectious diseases and human factors accelerating the development of AMR, it would be interesting to note the differences in the approaches and implementation issues of AMR between developed and developing countries. Furthermore, the Global Action Plan (GAP) proposes that member nations link their AMR National Action Plans (NAPs) with the GAP, which includes country-specific perspective. To have a reflection of the similarities and differences of National AMR plans to contain AMR and the implementation issues of both developed and developing countries with that Global Action plan, the authors have reviewed various published documents, other sources of information highlighting these issues. The NAPs of the countries selected for review are as follows:

1. Developing countries: Cambodia, Philippines, Indonesia, Laos, Myanmar, Thailand, Brunei, Vietnam, India, and Nigeria [4].
2. Developed countries: Australia, Canada, USA, South Africa, Switzerland, and Singapore.

8.2 Global initiatives for containment of AMR

A recent review of global AMR burden highlighted that the problem has reached critical levels with around 4 million deaths associated with AMR. The spread of AMR has no bars and thus, no nation can avoid the negative impact of misuse of antibiotics alone. Thus, there is need to coordinate the actions with the rest of the globe with a common endeavor to provide global public goods [5].

The movement started in 2001, when the World Health Organization (WHO) devised a framework for a worldwide strategy for AMR prevention [5]. The growing threat of a catastrophe prompted the adoption of the Global Action Plan (GAP) on AMR at the 67th World Health Assembly in 2015 [3]. It also attempted to establish and implement multisectoral National Action Plans (NAP). According to WHO, 100 countries have AMR national action plans in place, while 51 nations are in the process of developing AMR plans. [5] using One health approach [3]. In 2016, the need to bring together the allies from the United Nations, international organisations, and persons with experience in human, animal, and plant health, as well as food, animal feed, commerce, advancement, and the environment, were brought into the picture. In 2019, a Tripartite Joint Secretariat comprised of FAO, OIE, and WHO was established to encourage multi-stakeholder participation [6]. The GAP basically focuses on five strategies – awareness, surveillance, innovation, optimum use, and infection prevention. The awareness at global level is created through 'World Antimicrobial Awareness Week' (WAAW). It is a global initiative with the overriding motto 'Antimicrobials: Handle with Care,' with the goal of raising awareness of AMR globally and encouraging best practises among the general public, health workers, and policymakers. In 2015, WHO established the Global Antimicrobial Resistance and Use surveillance system (GLASS). GLASS has 92 nations, territories, and regions enrolled as of April 2020, including low-, middle-, and upper-income countries. Furthermore, the Global Antibiotic Research and Development Partnership (GARDP) focuses on the 5 by 25 approach. This essentially intends to produce 5 novel medicines that are effective at killing drug-resistant microorganisms by the year 2025 [7]. CDC is actively involved in all three aspects both at national and global level. It has made huge investment [7], notably in low- and middle-income countries to build the laboratory capacity to detect resistant infections and innovate drugs. The agency also develops guidelines on infection prevention and control with an aim to prevent infections in the first place. Social media and websites also play a key role in sharing relevant information on different aspects of AMR such as ReAct, Keep Antibiotics going Coalition, stopamr, preventit, amrinsights, etc. The best part is the voluntary commitment made by parties, to more robust legal-style commitments. One of the major chicken producers in the US, such as Perdue Farms and McDonalds has eliminated the use of antibiotics in their production [3].

8.3 Policy and Implementation issues of AMR in developing countries

The policy and approaches of AMR containment and implementation issues have been discussed from the viewpoints of policy perspective; implementation perspective; monitoring and evaluation perspective; and one health perspective largely adapted from the article by [4] reviewing AMR action plans in southeast Asia.

8.3.1 Policy perspective

Policy perspective is as follows.

8.3.1.1 Validity period and stakeholder involvement

In most of the NAPs, specific validity period has been assigned. To name a few, India, Nigeria, Vietnam, Cambodia, Indonesia, and Philippines have stated the validity period clearly. For designing and implementation of NAPs multiple sectors have been extensively involved. For example, in Vietnam, Ministry of Health, Ministry of Agriculture, and Rural Development; For India "Ministry of Health and Family Welfare, National centre for Disease control, National Institute of Pharmaceutical education and Research, Veterinary Council of India, Ministry of Human Resource Development, Ministry of Information and Broad casting, Consumer affairs food and consumer distribution," etc. have been involved [4].

8.3.1.2 Objectives of GAP and NAPs

AMR's Global Action Plan (GAP) contains five goals.: (1) to improve awareness and understanding of antimicrobial resistance through effective communication, education, and training; (2) to strengthen the knowledge and evidence base through surveillance and research; (3) to reduce the incidence of infection through effective sanitation, hygiene and infection prevention measures; (4) to optimize the use of antimicrobial medicines in human and animal health; (5) to develop the economic case for sustainable investment that takes into account the needs of all countries and to increase investment in new medicines, diagnostic tools, vaccines and other interventions (GLASS 2015). Most of the NAPs have four to seven objectives and majority of them are aligned with those of GAP (GLASS 2015). First four objectives of almost all the NAPs aligned with the GAP objectives. However there are certain variations. The 5th objective of NAP i.e., Establishing the economic rationale for long-term investment has received less attention.. Some NAPs have also included a 6th objective. For example, Philippines and Thailand, emphasised the need of having a steady supply of high-quality drugs as a 6th objective while a the sixth target, India emphasized'strengthening India's leadership on AMR, especially international, national, and state level cooperation.'[8][4].

8.3.1.3 Coordination and accountability

In almost all the NAPs, coordination between sectors do exist for a multisectoral approach to combat AMR. The need for intersectoral committee for coordination and implementation has been highlighted. Most of the NAPs have highlighted about "Technical Working Groups (TWGs)" to work with relation to particular topics. For example, in India, the "Intersectoral Coordination Committee" on AMR (ICC-AMR)(Ministry of Health and Family Welfare 2017) do exist consisting of secretaries and joint secretaries of various ministries and departments. The Intersectoral Coordination Committee on Antimicrobial Resistance in India's objective is to supervise and coordinate policy choices for antimicrobial resistance initiatives in all sectors in conformity with AMR-related public health goals. "Technical Working Groups (TWGs)" of India consisting of experts from the various ministries, government departments, institutions and national and international organizations [8]. Accountability aspects of the NAPs have not been adequately discussed in all NAPs. However, in some NAPs accountability and political commitment have been reflected. For example, in India, Vietnam, and Nigeria have stated political and multi-sectoral commitments in implementation of NAPs especially in the in the domains of surveillance, stewardship and one health approach to contain AMR [9] [4].

8.3.1.4 Equity

Accessibility and affordability to healthcare and quality antimicrobials are the two main equity factors in containment of AMR. In addition, gender and socio-economic considerations should also be considered as equity factors in the NAPs. Equity issues have been addressed in relation to accessibility and affordability to quality antimicrobials by most of the countries.

8.3.2 Implementation perspective

Implementation perspective highlights interventions to contain AMR. This has been discussed from the domains of awareness, education and training; surveillance; Development of Research and innovation activities in AMR; prevention and management of infections; optimization of antibiotic use in human and animal health, infection prevention and control, and international collaboration.

8.3.2.1 Awareness, Education and Training

Awareness of the stakeholders on AMR across multiple sectors (human, animal, agriculture, and food) have been one of the key priorities in almost all the NAPs. Inclusion of AMR curriculum in basic education including school education and professional education have been important in all NAPs. In addition, most of the NAPs have continuing professional education programmes, such as on-the-job training, workshops, and conferences, critical initiatives for expanding AMR knowledge. Public awareness on AMR has been featured in all the NAPs [8, 4].

8.3.2.2 Surveillance

In almost all NAPs, surveillance for antimicrobial use (AMU) and resistant organisms, and support for adequate laboratory capacity and capability both in human and animal sectors have been discussed. Very few countries have highlighted environmental AMR surveillance. Strengthening of existing surveillance system, and its further extension involvement of both private sectors have also been mentioned in most of the countries such as India, Vietnam, Nigeria, etc. [8, 4]. Most of the countries excepting few, i.e, Brunei and Philippines have cited the development of early identification and reporting strategies for newly developed resistance. In almost all NAPs, healthcare-associated infections (HAIs) surveillance has been highlighted. Most NAPs have emphasised regional and national surveillance networks, as well as integration for greater information exchange.

8.3.2.3 Development of research and innovation activities in AMR

Research and development and related innovation have been mentioned in all NAPs to combat AMR. Financial provision to encourage Research and development has also been mentioned in NAPs. Research areas such as: understanding the drivers of AMR; knowledge, attitude and practices of stakeholders; creation of evidence; multi-stakeholders research agenda; and the impact of AMR strategies have been mentioned in the country NAPs.

8.3.2.4 Infection prevention and control

Inadequate sanitation and hygiene practices facilitate in spreading the drug-resistant infections. The goal of infection prevention and control (IPC) is to minimise the spread of resistant organisms and illness hazards. Most of the NAPs, specifically in the human and animal health sectors, IPC policies and recommendations have been widely explored., setting of infection prevention and control committee in health facilities, and management of water, sanitation and hygiene. Good agricultural practices and biosecurity improvements have been proposed to prevent diseases in animal health sector. Countries namely India, Cambodia, Indonesia, Lao, Myanmar, and Nigeria have described Environmental IPC. In some countries like India, Cambodia, Nigeria, Myanmar, and Philippines have cited IPC for food products. Immunization programme as a strategy in reducing infections both in human and animal sectors has been cited by India, Nigeria, Vietnam, Cambodia, Indonesia, Malaysia, and Myanmar, etc [8, 4].

8.3.2.5 Optimising antimicrobial usage in human and animal health

Most of the developing countries have mentioned antimicrobial stewardship programmes (ASPs) in their NAPs both for human and animal sectors. In majority of NAPs, antimicrobial use policy to restrict irrational use of antimicrobials has been mentioned. In India, specific mention has been made to (1) Enhance national regulatory agencies for better antimicrobial quality, safety, and efficacy; and (2) Generate and implement an enabling regulatory framework and intersectoral cooperation for antimicrobial use in animals and food safety legislation (Ministry of Health and Family Welfare 2017). In some of the NAPs such as Philippines and Malaysia, rewards or sanctions for reducing excessive AMU have been suggested. The employment of enforcement agencies to support rules has been highlighted in various NAPs [4].

8.3.2.6 International collaboration

No country can combat AMR alone. It requires partnership and collaboration at local, national and international levels. All countries have discussed partnerships and engagement of donors, and development partners except Brunei in combating AMR [4]. Technical support and funding for activities to contain AMR have been highlighted in country NAPs. India has specifically mentioned its commitment and collaboration on AMR at international level in strategic priority 6 of India's NAP [8].

8.3.3 Monitoring and evaluation

Monitoring and Evaluation (MandE) have found a place in all the NAPs. Most of the countries have MandE indices either fully or partially. All NAPs have a mention about data reporting on AMR.

8.3.4 One Health perspective

All NAPs have designed their action plan on AMR mentioning one health approach. While emphasis has been given to human, animal sector, and environmental sector requires more elaboration.

8.4 Policy and implementation of AMR in developed countries

The policy and approaches of AMR containment and implementation issues have been discussed from the viewpoints of policy perspective; implementation perspective; monitoring and evaluation perspective and one health perspective.

8.4.1 Policy perspective

Policy perspective includes the following.

8.4.1.1 Validity period and stakeholder involvement

In most of the NAPs, specific validity period has been assigned and from time to time the NAPs are being revised. In formulation and implementation of NAPs multiple sectors cutting across human, animal, food, and agriculture sectors both in public, private and non-governmental sectors have been involved. For example, in Switzerland, the Federal Office of Public Health (FOPH), the Federal Food Safety and Veterinary Office (FSVO), the Federal Office for Agriculture (FOAG), the Federal Office for the Environment (FOEN), and the Swiss Conference of Cantonal Ministers of Public Health (CMPH), have been involved [10].

8.4.1.2 Objectives of GAP and NAPs

The NAPs of developed countries reviewed have four to eight objectives. Most of the strategic objectives of GAPs are mentioned NAPs of developed countries. In the NAP of South Africa, legislative and policy reform for health systems strengthening; education and workforce development; communication and research have mentioned as strategic enablers [11].

8.4.1.3 Coordination and accountability

In all the NAPs of developed countries reviewed, they have a clear mention about the coordination mechanism between different sectors to contain AMR. Inter-sectoral committee for coordination and implementation has been highlighted. Accountability aspects of the NAPs have also been mentioned [8, 12, 11, 10].

8.4.1.4 Equity

Accessibility and affordability to healthcare and quality antimicrobials are emphasized in the NAPs.

8.4.2 Implementation perspective

Implementation perspective highlights interventions to contain AMR. This has been discussed from the domains of awareness, education and training; surveillance; development of research and innovation activities in AMR; prevention and management of infections; optimization of antibiotic use in human and animal health, infection prevention and control, and international collaboration.

8.4.2.1 Awareness, Education and Training

AMR public education and professional education/training has been mentioned in all NAPs of developed countries reviewed by the authors, however the educational and training component has been placed under different goals/ objectives/strategic enablers[8, 12, 11, 10].

8.4.2.2 Surveillance

In almost all NAPs, surveillance for antimicrobial use; resistant pattern; laboratory surveillance; surveillance across one health; reporting to international database on AMR have been discussed. The NAP of South Africa mentions early warning systems for sentinel organisms and epidemics [8, 12, 11, 10].

8.4.2.3 Development of Research and innovation activities in AMR

Research and innovation have been mentioned in all NAPs. Countries have prioritised their research areas, many of which are common to the countries. Country wise identified research areas are as follows:

- Australia: AMR in environment; resistance transmission risk between human, animal and environment; transfer of research findings to grass root level across all sectors.

- Canada: Cross-sectional multidisciplinary research; best practices, behavioural research; economics and production impact across sectors; capacity development, alternative to antimicrobials, new diagnostic tools, incentivise pharmaceutical investment [13].

- Singapore: Transmission pathway, advance surveillance method, socio-behavioural research, prescription pattern, knowledge, attitude and practice and socio-economic impact on AMR, development of national research agenda [8, 12, 11, 10].

- South Africa: Point-of-care testing, antimicrobial prescribing and usage, behaviour modification, antimicrobial intake, and new antimicrobials are all examples of novel diagnostics [11]

- Switzerland: Interdisciplinary research on emergence, transmission, dissemination, and management of resistant bacteria; the focused production of antimicrobial compounds; and the development of cost-effective diagnostics products [10]

- USA: International collaboration, public private partnership; incentivization for the development of new antibiotics; alternates to antibiotics and low-cost point-of-care diagnostics [14]

8.4.2.4 Infection prevention and control

All NAPs have clearly mentioned the infection prevention and control (IPC) strategies. Country-wise strategies of IPC are as follows:

- Australia: Best practices in IPC, biosecurity programs and standards; quality assurance; monitoring and evaluation across all sectors; AMR stewardship and timely use of monitoring and surveillance data; new and better ways to arrest the spread of infection, and occurrence of critically important resistant infections [15].

- Canada: Communication, education and training programme

and tools on evidence based IPC practices; facilitate and promote application and oversight of IPC best practices; development of policy guidelines; stand setting and knowledge translation; capacity building; reducing inequalities in delivering comprehensive and effective IPC programmes in human and animal sectors; and invest in IPC research [13].

- Singapore: Food safety and hygiene; safe food handling education; surveillance of pathogens in food; policies and guidelines on IPC; autogenous vaccine in aquaculture; vaccination monitoring and management in farms; incentives for good husbandry practices; and good aquaculture practices. [9, 8].

- South Africa: Targeted campaigns for communities knowledge of infection prevention and proper antibiotic usage in both the human and animal health sectors; decreased use of antimicrobials through disease prevention and community initiatives such as vaccination programmes, water and sanitation improvements, and better biosafety [11].

- Switzerland: Improved hygiene; farm cleanliness and husbandry, as well as the discovery of viable antibiotic alternatives [10].

- USA: Assist nations in developing and implementing national antibiotic-resistance plans and strategies to improve antimicrobial stewardship; promoting antibiotic quality, safety, and efficacy and strengthening the pharmaceutical supply chain; and coordinating regulatory approaches through collaboration with international organisations [14]

8.4.2.5 Optimising antimicrobial usage in human and animal health

All the countries mentioned the strategies for optimising antimicrobial use in human and animal health. Some of strategies of different countries mentioned in the NAPs are as follows:

- Australia: Collection and integrated reporting of Antimicrobial Use and Resistance data by the country's surveillance system; Standards for and accreditation of veterinary practices and hospitals to support the implementation, maintenance and improvement of antimicrobial stewardship programs and best-practice in antimicrobial use [13]

- Canada: Partnership and collaboration between governments and relevant stakeholders for the integration of AMR data and information, and surveillance for strengthening the monitoring and reporting of antimicrobial use [8].

- Singapore: Antimicrobial stewardship programme in hospitals and veterinary care; antimicrobial use monitoring; hospital and clinic AMU guidelines; NAP mentions a veterinary medication registration system and supply chain supervision [8].

- South Africa: Development of an antimicrobial surveillance system; AMR surveillance system in food-producing animals; every institution should have an antimicrobial stewardship team to supervise appropriate prescription and optimise antibiotic use [8].

- Switzerland: Refining of roles on rational use of antibiotics based on current level of understanding [10].

- USA: Guidance from the Food and Drug Administration regarding the prudent use of medically essential antimicrobial medicines in food-producing animals. [14]

8.4.2.6 International collaboration

All countries have mentioned about the collaboration with international organizations. The organizations mentioned by different countries for collaborations are: United Nations, WHO, G20, G7, Tripartite agencies (the World Health Organization, the Food and Agriculture Organization of the United Nations, and the World Organisation for Animal Health); countries in South East Asia and Pacific regions; European union, etc [8, 12, 11, 10].

8.4.3 Monitoring and evaluation

Monitoring and Evaluation (M and E) strategies have been mentioned in each country's NAP. Emphasis has been given on reporting and feedback mechanisms. Country specific monitoring indices have developed [8, 12, 11, 10].

8.4.4 One Health perspective

The Global Action Plan on AMR highlights the need for using "one health" approach involving human, veterinary, agriculture, finance, environment sectors, and consumers [3]. All the developed countries selected for review have adopted the one health approach in their National AMR action plans.

8.5 Conclusion

No wonder variations were observed both in developed and developing countries to make it a more country specific AMR action plan. Developing countries are in different stages of implementing the national action plans. Inter country variations have been observed both in developing and developed countries in terms of number of objectives and strategies to achieve these objectives. Strategies to incentivise development of newer antibiotics, alternative to antibiotics have been prominently elaborated in NAPs of developed countries. Containment of spread of resistant microbes through environment requires more emphasis in developing countries. Intensive approach required for community awareness on AMR. Customised training programme for the drug dispensers specifically in developing countries should be given priorities to reduce over-the-counter access to antibiotics and AMR. Developing countries should focus more on research and development on AMR, evidence generation and international collaboration.

Bibliography

[1] World Health Organization et al. Who global strategy for containment of antimicrobial resistance. Technical report, World Health Organization, 2001.

[2] James A. Ayukekbong, Michel Ntemgwa, and Andrew N. Atabe. The threat of antimicrobial resistance in developing countries: Causes and control strategies. *Antimicrobial Resistance and Infection Control*, 6(1):1–8, may 2017.

[3] WHO GLASS. Global Antimicrobial Resistance and Use Surveillance System (GLASS), 2015.

[4] Alvin Qijia Chua, Monica Verma, Li Yang Hsu, and Helena Legido-Quigley. An analysis of national action plans on antimicrobial resistance in Southeast Asia using a governance framework approach. *The Lancet Regional Health - Western Pacific*, 7:100084, 2021.

[5] WHO. Antimicrobial resistance, 2001.

[6] World Health Organization. Antimicrobial resistance, 2021.

[7] CDC. CDC funds 34 innovative projects to combat antibiotic resistance — CDC Online Newsroom — CDC, 2016.

[8] National Water Agency (PUB) Agri-Food Veterinary Authority of Singapore (AVA), Ministry of Health (MOH), National Environment Agency (NEA). Singapore: National strategic action plan on antimicrobial resistance, 2017.

[9] Environment Federal Ministries of Agriculture and Health. Nigeria AMR Action Plan. 2017.

[10] The Federal Council. Strategy on Antibiotic Resistance Switzerland. page 74, 2015.

[11] Forestry Departments of Health and Agriculture and Fisheries for the Republic of South Africa. South Africa: South African antimicrobial resistance national strategy framework, 2018.

[12] Government of India. National action plan antimicrobial resistance (nap-amr) 2017–2021. 2017.

[13] The Canadian Committee on Antibiotic Resistance. Infection Prevention and Control Best Practices, 2008.

[14] United States. NATIONAL ACTION ANTIBIOTIC-RESISTANT BACTERIA Progress Report : Year 3. (October), 2018.

[15] Joan A Casey, Brent F Kim, Jesper Larsen, Lance B Price, and Keeve E Nachman. Industrial food animal production and community health. *Current environmental health reports*, 2(3):259–271, 2015.

9

Implementation Challenges in Healthcare-associated Antimicrobial Resistance Prevention and Control in India

Prakash Narayanan
Prasanna School of Public Health, Manipal Academy of Higher Education, Manipal, Karnataka, 576104, India

Rakshitha K.
Prasanna School of Public Health, Manipal Academy of Higher Education, Manipal, Karnataka, 576104, India

Helmut Brand
Department of International Health, Care and Public Health Research Institute – CAPHRI, Maastricht University, Maastricht, The Netherlands

CONTENTS

9.1 Introduction

One route of transmission of drug resistant pathogens to human beings is the healthcare setting. The presence and abundance of infected and drug-resistant pathogens in healthcare settings often convert such settings as the source of transmission of infectious diseases and drug resistance to healthcare professionals, patients and the visitors in those settings. Healthcare associated infection (HAI) or 'nosocomial infection,' is "an infection occurs to the patient in a hospital or other health care facility in whom the infection was not present or incubating at the time of admission". This includes occupational infections among staff of that facility and the infections that are gained by patients in the hospital but appearing after discharge (WHO no date). Improper use of invasive devices such as ventilators and catheters, used in modern healthcare, also causes these infections. Incorrect methods used for the collection and handling of human samples for laboratory investigations in hospitals often cause the spread of these infections to the sample collectors, laboratory staffs, and patients. Inappropriate use of personal protection equipment and hygienic practices of staffs, patients, and visitors increases the risk for such infections including antimicrobial resistance. Understanding the challenges in implementing infection prevention and control activities for AMR help in devising appropriate cost-effective and sustainable strategies suitable to institutions.

9.1.1 Types of hospital-acquired infections

Modern healthcare setting employs invasive procedures and devices in treating patients. The patients, those accompanying the patients and those who work in the hospitals are susceptible to many infections from healthcare settings. Most common types of HAI are as follows [1].

9.1.1.1 Ventilator-associated pneumonia

Ventilator-associated pneumonia is a serious lung infection that affects patients on ventilator support, which develops 48 hours or more after endotracheal intubation and not appeared to be incubating at the time of admission [2]. It makes clinical suspension with manifestations such as fever, increase in white blood cell count, purulent tracheo-bronchial secretion along with new infiltrates present on chest X-ray. Usually confirmed with positive blood culture for the same pathogen found in lower respiratory tract secretions. The antibiotic-resistant strains are of important concern here, more likely pathogens differ from the community-acquired pneumonia [3].

9.1.1.2 Central line-associated bloodstream infection (CLABSI)

Vascular catheters are placed for intravascular treatments. CLABSI is laboratory-confirmed bloodstream infection, in the patient who had a central line at the time of infection or within 48 hours before development of infection and not related to an infection on another site [4].

9.1.1.3 Catheter-associated urinary tract infection (CAUTI)

Nosocomial infections are associated with urinary tract following the catheter instrumentation. It is caused when the urinary catheter is not inserted properly, not cleaned and left for too long unchanged, infection can be noted in organs of the urinary tract. Catheters act as channels for entry of bacteria whereas the imperfect drain retains some volume of urine in the bladder, posing as bacterial residence [4].

9.1.1.4 Surgical site infection (SSI)

According to Centers for Disease Control and Prevention (CDC), SSI is the infection that occurs at an incision site within 30 days of surgery. Contamination from the external environment and poor hospital hygiene maintenance are probable reasons for higher rates of infection [5].

9.2 Burden of HAI

HAIs are a global crisis affecting millions of patients and health-care workers every year. The burden of HAIs is significantly higher in lower middle-income countries than other countries (WHO No date). Many developed countries, such as the United States (US) and European Economic Areas have well established HAI surveillance systems for collecting data on health, economic burden, and patient care.

9.2.1 HAI in developed countries

The CDC managed National Healthcare Safety Network is the largest and mostly used electronic surveillance system for tracking HAIs in the US. Over 25000 healthcare facilities use this network and data is used to analyze HAIs, adverse events and healthcare personnel safety. Most frequently reported HAI causing pathogen across the healthcare facilities were *Escherichia coli* (18%). *Staphylococcus aureus* (12%) and *Klebsiella* spp (9%) [6]. As per the report published in 2018 by European Centre for Disease Prevention and Control (ECDC), the point prevalence survey of 2016-17 estimated that 6.5% patients of acute care hospitals (ACH) and 3.9% residents in long-term care facilities (LTCF) in European Union and European economic area (EU/ EEA) had at least one HAI. The number of episodes of HAIs per year was 8.9 million. Antimicrobial resistance (AMR) to selected AMR markers was 31.6% in ACH and 28.0% in LTCF [7] For the period of 2015-19, AMR in European Union reports *E. coli* (44.2%) as the most commonly noted bacteria among the bacterial species under surveillance. Followed by *Staphylococcus aureus* (20.6%). Increase in the percentage of Vancomycin-resistant isolates of *Enterococcus faecium* from 10.5% in 2015 to 18.3% in 2019 was a particular concern. Resistant percent was high in *Klebsiella pneumoniae,* where several countries reported resistance above 10% [8]).

TABLE 9.1
Device associated HAI rates, excess duration of hospital stays and mortality in India 2004-2013

Type of DA-HAIs	Pooled device associate HAI rate	Length of extra hospital stay (days)	Crude extra mortality
Ventilator-associated pneumonia (VAP)	9.4 per 1000 ventilator days	9.1	22.7%
Central line-associated bloodstream infection (CLABSI)	5.1 per 1000 catheter days	9.5	16.3%
Catheter-associated urinary tract infection (CAUTI)	2.1 per 1000 catheter days	10	6.6%

9.2.2 Burden of HAI in developing countries

Most of the developing countries lack surveillance systems of HAIs, and often struggle with the complexity and lack of standardized criteria for diagnosing the infections (WHO Nodate). The limited data, usually of inferior quality, makes it difficult to gather information at the global level. A systematic review and meta-analysis of burden of HAI in Southeast Asia estimated the pooled prevalence of overall HAIs was 9.0%, pooled incident density was 20 cases per 1000 intensive care unit days [8]. The excess length of stay because of HAI was between 5-21 days. The attributed mortality ranged from 7% to 46% [9].

The Device Associated- Healthcare Associated Infections are the primary cause of patient mortality and morbidity and are the principal threats to patient safety. The pooled rates of device-associated HAIs in India are higher than the higher income countries. A study conducted by International Nosocomial Infection Control Consortium across 20 cities in India for 10 years, included 84 ICUs in 40 hospitals reported the pooled DA-HAIs for adult and paediatric ICUs as shown in Table 9.1 [10].

The knowledge of the true burden of antimicrobial resistance in HAIs is required to guide effective empirical therapy. In India,

TABLE 9.2
Common organisms, resistant mechanisms, and resistant drugs

Organisms	Some resistant mechanisms	Resistant drugs
Gram-positive organisms		
Staphylococcus aureus	Aminoglycoside-modifying enzymes (AMEs)	Penicillin related drugs Gentamicin Erythromycin Co-trimoxazole Clindamycin Vancomycin Daptomycin
Streptococcus pneumoniae	a. modified penicillin-binding proteins (PBPs)-altered target site b. Mobile genetic elements *mef* (A/E)	Fluoroquinolone Penicillin Erythromycin
Gram-negative organisms		
Escherichia coli	a. production of extended-spectrum β-lactamase enzymes (ESBL) b. plasmid-mediated resistance	Penicillins Cephalosporins Carbapenems Polymyxins Tetracycline Trimethoprim Sulphonamides Chloramphenicol Streptomycin Colistin
Klebsiella pneumoniae	a. ESBL b. AME c. chromosomal mutations	Carbapenem Colistin Tigecycline
Acinetobacter baumannii	a. β-lactam resistance b. AME c. Gene mutation	Class D Carbapenemase Aminoglycoside antibiotics Colistin

there is no systemic or national surveillance program. Individual healthcare settings and a few multicentric studies have reported the data on AMR (Table 9.2). In 2014, Indian Council of Medical Research (ICMR) established an antimicrobial resistance surveillance network in collaboration with tertiary care hospitals across the country [11]. As per the review published in 2019, the higher rates of resistance in India were seen in Gram-negative organisms than in Gram-positives [12].

9.3 Causes/ Risk factors associated with Hospital Acquired infections

The inappropriate healthcare practices cause the Hospital Acquired Infections in hospitals (Table 9.3.). The risk factors that contribute to higher hospital acquired infections are given below:

1. Improper maintenance of sterile and hygiene conditions: Interventions and surgeries are an integral part of the modern medicine system. Usage of devices such as urinary catheters, ventilators, central lines, procedures such as endotracheal intubation if performed in unhygienic ways can lead to hospital acquired infections.

2. Contaminated instruments: Appropriate sterilization of instruments is important to prevent hospital acquired infection.

3. Congested hospital setting: overcrowding in the hospital settings can lead spread of infections among patients and hospital staffs (e.g., placing the hospital beds in close proximity)

4. Contaminated air conditioning can risk the safety of patients.

5. Under staffing: Hospitals may fail to give necessary care to the patients if the human resource is less.

6. Improper antimicrobial usage: Antibiotics are the chief mea-

TABLE 9.3
Risk factors by type of healthcare associated infections [9]

Type of HAI	Risk/influencing factors
VAP	Longer duration of mechanical ventilation Cancer Leukocytosis Aspiration pneumonia
CLABSI	Duration of infusion of CVC or intravenous catheters Frequency of change of CVC Clinical sepsis Methicillin- resistant *Staphylococcus aurus*/in axilla/perineum/throat Use of hydrocortisone
UTI	Indwelling urinary catheter and frequency of change
SSI	Duration of surgery Degree of wound contamination Small bowel surgery Preoperative stay Emergency operations

sure in clinical treatment. The irrational usage can lead to recurrent infections and development of resistant strains.

7. Carelessness towards vulnerable populations: Patient-related factors also play an important role in acquiring and transmission of hospital acquired infections. The infants with an immature immune system and elderly persons (greater than 60 years old) with low immune functions are more susceptible for hospital acquired infections. Patients with immuno-suppression disorders and individuals involved in invasive surgeries are also vulnerable.

9.4 Spread of healthcare-associated infections

Hospitals are the source of the microbes that cause HAIs. The spread of nosocomial infections can be of two ways, direct and indirect, as shown in Figure 9.1 [13]. The contaminated hands are the principal route of disease transmission. The direct contact spread of infection is by touching the infected individuals or any reservoir of infection. This can be in following ways: staff to patient; patient to visitor; patient to patient and visitor to patients. Inappropriate use of personal protective equipment can cause acquisition and transmission of such infections. The indirect contact involves aerosols, vectors and usage of contaminated instruments. Aerosol spread occurs when the infected patient or care provider coughs or sneezes. The respiratory tract infections mainly transmit through this mode. *Pseudomonas, Acinetobacter baumannii, Legionella,* and *mycobacterium* are some bacteria that can transmit through air. Vectors are involved in transmission of infection by dispersion of causative microbes. The flies, ants, cockroaches come into direct contact with contaminated surfaces or excreta of infected patients and are involved in cross contamination to transmit the infection. *Klebsiella pneumonia, Staphylococcus aureus, Salmonella, Pseudomonas aeruginosa, Escherichia coli* are a few examples. More than 80 bacterial species including the resistant strains have been identified [14].

9.5 Detection of AMR- HAI

Classical phenotypic antibiotic susceptibility testing (AST) refers to the broth microdilution (BMD) or diffusion. The Clinical and Laboratory Standards Institute and the European Committee on Antimicrobial Susceptibility Testing both consider BMD to be the gold standard for Minimum Inhibitory Concentration/ culturing microorganisms breakpoint determination [15]. The AST profiling can take up to one week after preliminary diagnosis. This delay can

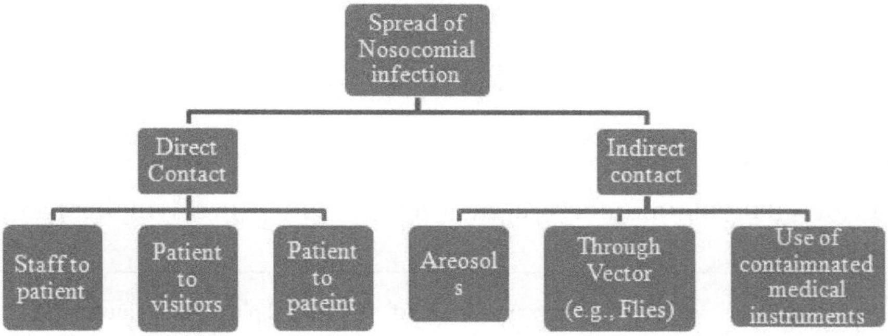

FIGURE 9.1
Modes of transmission of Nosocomial infections [14]

be lethal for a few patients, so many rapid AST diagnostic methods came in place with their own advantages and disadvantages [16].

9.6 Infection Prevention and Control Policy

The problem of antibiotic resistance in healthcare associated infections is multifaceted and catastrophic as it includes patients, health care providers, pharmaceutical industries, policy makers and the public. Evidence has shown that Infection Prevention and Control (IPC) is a scientific approach to prevent HAIs causing harm to patients and healthcare workers. Infection control measures are crucial in prevention of spread of diseases. Implementation of hospital infection prevention and control programmes can reduce mortality and morbidity among patient's and hospital staffs. IPC can also considerably reduce the length of hospital stay and expenditure associated with it.

9.6.1 Prevention of HAIs

The patients are exposed to different sources of pathogens during hospitalization such as healthcare environments, staff and other patients. HAI being a cause of illness, economic burden and death, need to be prevented from the source to control the spread (Table

TABLE 9.4
Major element of infection prevention and control in healthcare settings [17]

Organization of infection control at hospital level	It comprises nursing staff, a dedicated physician trained in IPC, microbiological facilities, data management support.
Ward occupancy and workload	Improved nurse-to-patient ratio, a higher number of permanent frontline healthcare workers can reduce HAI. High workload, long working hours and nurses working in different wards are potential risks for bloodstream infections.
Materials equipment and ergonomics	Hand hygiene facility at the point of care, easy to use pocket hand-rub, electronic reminders, customized insertion kits for central venous catheters helps to decrease rate of HAI
Use of guidelines, education, and training	National guidelines and attitude of healthcare staffs to it
Team oriented and task-oriented education and training	Simulation based training, conducting workshops for physicians who are training the staff, multidisciplinary focus groups can improve adherence to IPC protocol.
Antimicrobial stewardship	Optimizing the use of antibiotics to treat bacterial infections. It can improve clinical outcome and prevent the risk of ABR
Standardization of audit	Auditing and personal feedback can improve predefined indicators for catheter insertion. Auditing the adherence to bundle care approach can prevent ventilator-associated pneumonia. Usage of comprehensive checklists which covers a wider range of healthcare practices can reduce all causes of HAI.
Prospective surveillance, feedback, and network	Participation in surveillance network and providing feedback to healthcare workers can reduce incidence of HAI

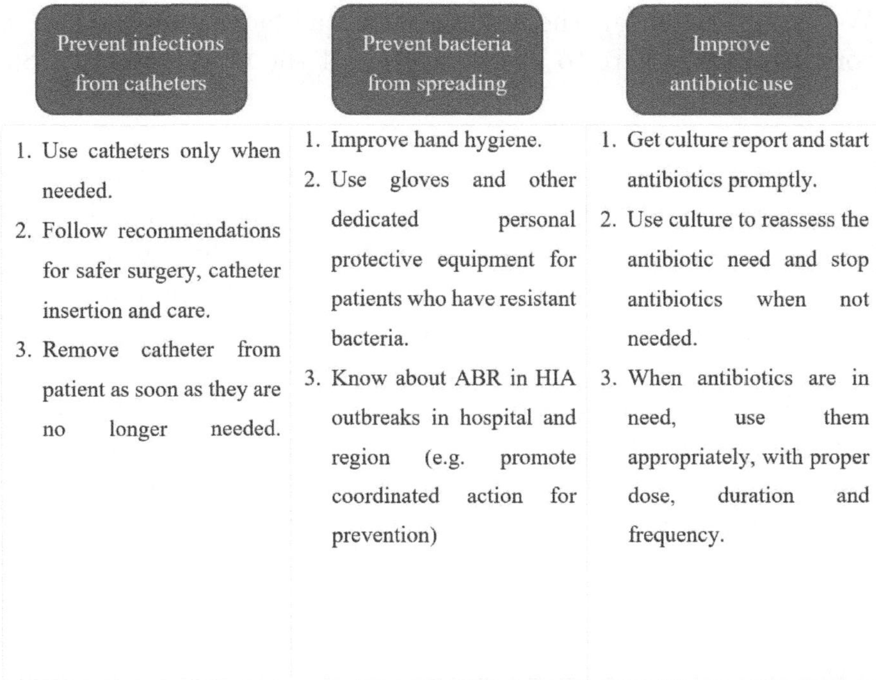

FIGURE 9.2
CDC recommended actions to prevent ABR infections in health-care settings [1]

9.4.). Antibiotic stewardship is to improve rational use of antibiotics in the healthcare settings and can enhance the IPC practices. Below is an example of how antibiotics are to be prescribed judiciously, a core component of AS explained (Figure 9.2.).

9.7 IPC core components

WHO identified eight core components that have to be established in countries to ensure an effective IPC programme. In the context, WHO also provides checklist with indicators for accountability and transparent health care [13]. The components are multidisciplinary, interrelated, and applicable to most healthcare settings.

We can establish all these components at facility level while six core components are to be established at the national level [18].

1. IPC programme- In each acute healthcare facility/healthcare level there should be IPC programme with a dedicated and trained team for preventing HAI and to combat AMR through IPC good practices. At the national level, stand-alone IPC programmes to be established for the same purpose.

2. IPC guidelines- Evidence based IPC guidelines to be developed to prevent HAI and ABR. To ensure successful implementation, the education and training of the workers to be conducted and monitoring of adherence with guideline recommendations should be undertaken.

3. IPC education and training- To limit the risk of HAI and AMR, all healthcare personnel should get IPC education through team- and task-based participatory strategies that include bedside and simulation training. At the national level, the programme should have the education and training as one of its core functions.

4. Surveillance- Facility-based HAI surveillance, including AMR surveillance with timely feedback to health care workers and stakeholders, as well as through national networks, should be conducted to guide IPC activities and detect outbreaks. National HAI surveillance programmes and networks should be established, with procedures for fast data feedback.

5. Multimodal strategies- The IPC activities to be implemented having multimodal strategies to improve good practices, reduce AMR and HAI. National IPC programmes should use multimodal methodologies to coordinate and facilitate the implementation of IPC activities on a national or subnational level.

6. Monitoring/audit of IPC practices and feedback- The timely feedback of healthcare practices as per the IPC standards should be recorded and promoted. At the national level, National IPC

monitoring and evaluation to be established.

7. Workload, staffing, and bed occupancy- Few of the elements such as bed occupancy, human resource/ workers staffing level should be adequate to reduce spread of AMR and HAI.

8. Build environment, materials, and equipment for IPC at the facility level- Required water, sanitation and hygiene facilities, hospital infrastructure to be in place to prevent infections.

9.8 HAI Management- Infection control bundles

Care bundles in infection prevention are set of evidence-based practices, generally three to five, when implemented collectively can improve healthcare delivery and patient outcomes [19] . These are used widely across healthcare settings to prevent and manage different health conditions. Unless medically contradicted every eligible patient should receive all the elements, the fidelity with care bundles should be 95% [20].

9.8.1 Benefits of care bundle approach

• It can ensure the delivery of minimum standard healthcare
• Bundles can be used as an audit tool to assess the delivery any intervention as a performance measure in healthcare setting
• Shorten hospital stay and reduces financial cost
• Engender hospital staff education in best patient care and promotes teamwork
• Improve resource utilization and multi-disciplinary collaboration
• It encourages the review of existing evidence and appropriate modification of clinical guidelines

To achieve patient protection, quality treatment, reduction in AMR and HAI cases, it is important to focus on infection prevention strategies. Clean and safe care is every patient's right and is a responsibility of all the healthcare working staff. The World Health Organization has issued specifications and recommendations in this regard to promote the minimum requirements of IPC in national and healthcare facility level. The minimum requirements are defined as, "IPC standards that should be in place at the national and facility level to provide minimum protection and safety to patients, healthcare workers and visitors, based on the WHO core components for IPC programmes" [13].

9.8.2 Role of hospital infection control committee

Infection prevention and control is a practical, evidence-based approach that prevents patients, visitors and healthcare workers from being harmed by avoidable and preventable infections. The IPC professional is defined by WHO as a "health care professional (medical doctor, nurse or other health-related professional) who has completed a certified postgraduate IPC training course, or a nationally or internationally recognized postgraduate course on IPC, or another core discipline including IPC as a core part of the curriculum and IPC practical and clinical training." The team of hospital infection control committee could contain Medical Superintendent Hospital administration; Consultant from microbiology, surgery, medicine, paediatric surgery, hospital administration and Nursing superintendent along with a few infection control nurses [13].

9.9 IPC in Indian context

In Indian context, the nation-wide initiatives where IPC was a strategic priority are, Biomedical waste management rule in 1998 (revised in 2016 and 2018), 'Kayakalp' programme in 2015, National guidelines on clean hospitals in 2015, ICMR released Guidelines for infection prevention and control in 2016, National quality health standards for public health facilities in 2017, National patient safety implementation framework 2018-2025. These show that infection prevention is at the core of public health in India, and the government is proactive to regulate the antimicrobial use. The Ministry of Health and Family Welfare of India released the National guidelines for infection prevention and control in healthcare facilities in 2020 [21]. The document includes having proper IPC guidelines, training and education materials, surveillance, multimodal strategies, monitoring audit and feedback as pillars of IPC programme and enabling environment as base. These new National Guidelines for IPC in Healthcare Facilities will improve patient safety and health workers' ability to prevent and manage infections in Indian hospitals, and also enhance overall quality of care.

9.9.1 IPC implementation challenges

The healthcare-associated infections are a major global problem in all healthcare systems, causing considerable mortality and morbidity to patients, hospital staffs, and visitors, and result in increased hospitalization costs [22]. Implementation of IPC in healthcare settings can reduce the risk of HAIs. The IPC guidelines and its implementation in low-and-middle-income countries vary by type of IPC programme, its robustness, and the surveillance systems they adopted. Implementation of IPCs is often resource intensive – budget constraints, lack of qualified staff, conflicting political agenda, and lack of surveillance. Medical standards and human capital are the key components of hospital accreditation, which stimulate continuous improvement and are not mandatory in India. As accreditation raises quality care and community confidence and hence the healthcare industry is turning towards it. There is an obvious interest among healthcare providers to improve IPC, and various healthcare settings have started infection control committees with varying standards. Despite these initiatives from Government of India, often proper implementation of IPC in Indian healthcare settings poses multiple challenges. We tried to summarize major IPC implementation challenges in the table below (Table 9.5.).

TABLE 9.5
IPC implementation challenges

Resource (human & infrastructure)	Organization/environment	Tools and technology
• Poor knowledge and application of basic infection control measures • Lack of knowledge of injection and blood transfusion safety • Lack of trained personnel in IPC • Inappropriate IPC training • Lack of education and training in prescribing antibiotics • Lack of clinical experience • Low hospital staff to patient ratio • High turnover of IPC trained staffs • Lack of infrastructure to implement IPC	• Lack of administrative support • Non-existence of hospital infection control committee • Heavy workload • Hospital overcrowding • Insufficient funding for IPC • Lack of IPC supplies • Lack of procedures for IPC • Insufficient equipment and suitable laboratories to test HAI/AMR • Non-existence of IPC and/or antimicrobial stewardship programme • Improper medical waste disposal • Non-mandatory medical standards and accreditation • Immuno-suppression and other severe underlying patient conditions • Insufficient application of standard and isolation precautions	• Comprehensive electronic health record • Adherence to personal protective equipment (PPE) • Surveillance system • Inadequate Laboratory facilities and diagnostic tools • Absence of easily accessible, open educational materials, and point of care tools • Language barrier in training materials • Lack of data sharing within and between institutions • Lack of monitoring and evaluation of IPC activities

9.10 Suggestions for effective implementation of HAI/ABR related IPC in healthcare settings

The effective infection prevention and control does not mean to have costly assets, and often can be implemented with minimum cost. Less costly interventions are feasible by adopting tactical approach, administrative will, sound practical skills, and clinical and administrative leadership [23].

According to WHO, the solutions to improve IPC for HAIs include (WHO Nodate):

- Identifying local determinants of the HCAI burden;
- Improving reporting and surveillance systems at the national level;
- Ensuring minimum requirements in terms of facilities and dedicated resources available for HCAI surveillance at the institutional level, including microbiology laboratories' capacity;
- Ensuring that core components for infection control are in place at the national and health-care setting levels;
- Implementing standard precautions, particularly best hand hygiene practices at the bedside;
- Improving staff education and accountability;
- Conducting research to adapt and validate surveillance protocols; and
- Conducting research on the potential involvement of patients and their families in HCAI reporting and control improving reporting and surveillance systems at the national level.

Healthcare institutions can device or adopt infection prevention and control programmes as per requirements. The hospital administration, healthcare workers, inpatients and visitors must take into such programmes, and need to play their roles for prevention and control of HAIs. The standard IPC practices such as hand and respiratory hygiene, PPE use, injection safety, storage and handling of medications, cleaning and disinfecting devices and environment, and medical waste management ideally not need to require extra regulations or resources. These standard procedures will protect persons from acquiring HAIs, including AMR. Other precautionary practices such as contact, droplet and airborne infections can be minimized to a great extent with the help of PPEs, personal hygiene and sanitation. Timely vaccination against vaccine-preventable diseases of persons working and visiting hospitals minimizes acquisition of many infectious diseases, and thus can minimize antimicrobial medication use. We can improve the implementation along with evaluation efforts by making the educational resources more simplified, free access and availability, and by incorporating graded recommendations from basics through progressive levels of IPCs and resources. The following cost-effective strategies are readily available and proven to prevent HAIs in resource-constrained settings, such as that exists in India [24].

- Presence of an IPC programme in healthcare setting and infection surveillance system
- Pertinent health care infrastructure
- Personnel trained in Infection control and working in appropriate number
- Continuous supply of antiseptics and disinfectants
- Regulated and timely carrying out sterilization procedures
- Promotion of hand hygiene among hospital staffs
- Vascular supplies and safe intravascular access practices
- Essential antimicrobials for treating infections
- Assurance of appropriate chlorinated water in the entire healthcare facility
- Political commitment, support of institutional policies and organizational structure from the hospital authorities.

Inclusion of healthcare-associated infection and infection prevention control in the academic curricula will help to build essential knowledge and skills to manage HAIs among new generation of healthcare professionals. Department of Health, Private hospital networks, Associations of healthcare professionals and Non-governmental organizations can develop, train and/or distribute no/low cost, easily comprehendible, and on the job materials and demonstrations to prevent and control HAIs to older generation of workforce. These organizations can be engaged to provide continued professional education on new technologies and solutions, and supportive supervision of IPC activities. Despite clear evidence of IPC efficacy, additional research is required to determine and validate innovative methods, cost effectiveness of IPC innovations and interventions, deciding on suitable implementation approaches, and local solutions. Periodical assessment and consultations with the implementors are vital for establishing and implementing sustainable healthcare associated infection prevention and control programmes.

Bibliography

[1] CDC. CDC funds 34 innovative projects to combat antibiotic resistance — CDC Online Newsroom — CDC, 2016.

[2] Alberto F Monegro, Vijayadershan Muppidi, and Hariharan Regunath. Hospital acquired infections. In *StatPearls [Internet]*. StatPearls Publishing, 2020.

[3] Andrea Coppadoro, EA Bittner, and Lorenzo Berra. Novel preventive strategies for ventilator-associated pneumonia. *Annual Update in Intensive Care and Emergency Medicine 2012*, pages 289–298, 2012.

[4] Yazan Haddadin, Pavan Annamaraju, and Hariharan Regunath. Central line associated blood stream infections. 2017.

[5] Raskin Chaudhary, Shrawan Kumar Thapa, Jid Chani Rana, and Pradeep Kumar Shah. Surgical site infections and antimicrobial resistance pattern. *Journal of Nepal Health Research Council*, 15(2):120–123, 2017.

[6] Lindsey M Weiner-Lastinger, Sheila Abner, Jonathan R Edwards, Alexander J Kallen, Maria Karlsson, Shelley S Magill, Daniel Pollock, Isaac See, Minn M Soe, Maroya S Walters, et al. Antimicrobial-resistant pathogens associated with adult healthcare-associated infections: summary of data reported to the national healthcare safety network, 2015–2017. *Infection Control & Hospital Epidemiology*, 41(1):1–18, 2020.

[7] Carl Suetens, Katrien Latour, Tommi Kärki, Enrico Ricchizzi, Pete Kinross, Maria Luisa Moro, Béatrice Jans, Susan Hopkins, Sonja Hansen, Outi Lyytikäinen, et al. Prevalence of healthcare-associated infections, estimated incidence and composite antimicrobial resistance index in acute care

hospitals and long-term care facilities: results from two european point prevalence surveys, 2016 to 2017. *Eurosurveillance*, 23(46):1800516, 2018.

[8] ECDC. Antimicrobial resistance in the eu/eea (ears-net)—annual epidemiological report for 2019, 2020.

[9] Moi Lin Ling, Anucha Apisarnthanarak, and Gilbert Madriaga. The burden of healthcare-associated infections in southeast asia: a systematic literature review and meta-analysis. *Clinical Infectious Diseases*, 60(11):1690–1699, 2015.

[10] Yatin Mehta, Namita Jaggi, Victor Daniel Rosenthal, Maithili Kavathekar, Asmita Sakle, Nita Munshi, Murali Chakravarthy, Subhash Kumar Todi, Narinder Saini, Camilla Rodrigues, et al. Device-associated infection rates in 20 cities of india, data summary for 2004–2013: findings of the international nosocomial infection control consortium. *infection control & hospital epidemiology*, 37(2):172–181, 2016.

[11] ICMR. Antimicrobial Stewardship Program Guideline, 2018.

[12] Balaji Veeraraghavan and Kamini Walia. Antimicrobial susceptibility profile & resistance mechanisms of global antimicrobial resistance surveillance system (glass) priority pathogens from india. *The Indian journal of medical research*, 149(2):87, 2019.

[13] WHO. Antimicrobial resistance, 2019.

[14] T Nivethithan and James D Raj. Endodontic pain-cause and management: A review. *International Journal of Pharmaceutical Sciences and Research*, 6(7):2723, 2015.

[15] David JP Mallon, John E Corkill, Sarah M Hazel, J Sian Wilson, Nigel P French, Malcolm Bennett, and C Anthony Hart. Excretion of vancomycin-resistant enterococci by wild mammals. *Emerging infectious diseases*, 8(6):636, 2002.

[16] Ekaterina Avershina, Valeria Shapovalova, and German Shipulin. Fighting antibiotic resistance in hospital-acquired infections: Current state and emerging technologies in disease prevention, diagnostics and therapy. *Frontiers in microbiology*, page 2044, 2021.

[17] Walter Zingg, Alison Holmes, Markus Dettenkofer, Tim Goetting, Federica Secci, Lauren Clack, Benedetta Allegranzi, Anna-Pelagia Magiorakos, Didier Pittet, et al. Hospital organisation, management, and structure for prevention of health-care-associated infection: a systematic review and expert consensus. *The Lancet Infectious Diseases*, 15(2):212–224, 2015.

[18] Julie Storr, Anthony Twyman, Walter Zingg, Nizam Damani, Claire Kilpatrick, Jacqui Reilly, Lesley Price, Matthias Egger, M Lindsay Grayson, Edward Kelley, et al. Core components for effective infection prevention and control programmes: new who evidence-based recommendations. *Antimicrobial Resistance & Infection Control*, 6(1):1–18, 2017.

[19] Deborah L Horner and Mark C Bellamy. Care bundles in intensive care. *Continuing education in anaesthesia, critical care & pain*, 12(4):199–202, 2012.

[20] Jacqueline F Lavallée, Trish A Gray, Jo Dumville, Wanda Russell, and Nicky Cullum. The effects of care bundles on patient outcomes: a systematic review and meta-analysis. *Implementation Science*, 12(1):1–13, 2017.

[21] WHO. Antibiotic resistance, jul 2020.

[22] Sangeeta Sastry, Nadia Masroor, Gonzalo Bearman, Rana Hajjeh, Alison Holmes, Ziad Memish, Britta Lassmann, Didier Pittet, Fiona Macnab, Rachel Kamau, et al. The 17th international congress on infectious diseases workshop on developing infection prevention and control resources for low-and middle-income countries. *International Journal of Infectious Diseases*, 57:138–143, 2017.

[23] World Health Organization. Antimicrobial Stewardship Programme in Health-Care Facilities in Low and Middle Income Countries, 2019.

[24] Robin van Kessel, Brian Li Han Wong, Ivan Rubinić, Ella O'Nuallain, and Katarzyna Czabanowska. Is europe prepared to go digital? making the case for developing digital capacity: an exploratory analysis of eurostat survey data. *PLOS Digital Health*, 1(2):e0000013, 2022.

10

Combating AMR through Behavior Change:Role of Higher Education Institutes

Meenakshi Sood

Chitkara School of Health Sciences, Chitkara University, Punjab

Vinay Kumari

Chitkara University College of Nursing, Chitkara University, Himachal Pradesh

Bunty Sharma

Department of Biotechnology, Maharishi Markandeshwar (Deemed to be) University, Mullana, Ambala

CONTENTS

10.1 Introduction

Infectious diseases have been affecting both humans and animals for eons and their treatment has always posed a problem. Discovery of anti-microbial drugs was a big breakthrough in this area as the new drugs saved a number of lives and decreased both morbidity and mortality. As new microorganisms resistant to commonly used drugs came into existence, this lead to a serious threat to health (Wright 2012). Antimicrobial resistance (AMR) is a diverse problem which affects not only humans and other living beings but also the environment[1].

It is a significant health issue of the twenty-first century that challenges the effectual prevention and management of a constantly escalating variety of infections due to microorganisms, no more sensitive to the usual antimicrobials employed to deal with them [2].

10.2 Burden of AMR

AMR has impacted at all levels, national and international. Statistically speaking approximately 0.7 million people die per year due to illnesses which are resistant to drugs. In the United States alone about 2.8 million humans get an antibiotic-resistant infection every year; and more than thirty-five thousand succumb as a result of it [3]. These resistant infections need longer hospitalization and expensive drugs which may be more toxic as well. Resistant infections also affect animal health resulting in challenges in disease prevention and treatment choices [4].

Antimicrobial resistance occurs when anti-infective drugs are ineffective and can neither kill nor stop reproduction of microbes.The factors responsible for AMR include the misuse and disuse of antibiotics, the transmission of infections to communities. Besides this the non-compliance to infection prevention and control procedures and spread of micro-organisms between animals and human beings via the food chain and various environmental factors such as improper disposal of wastewater also causes AMR [5, 6].

10.3 Requirement of Global Action

Four main areas are drawn in the development of antibiotic resistance: mankind treatment, animal production, agriculture, and the environmental section. The self-medication and inappropriate /irrational prescription by general practitioners consisting of over-prescription, inadequate dosing contributes to boosting antibiotic resistance. Practitioners with inadequate knowledge about antibiotics regarding its correct dosage, adverse effects and their perception of antibiotics as "wonder drugs" effective for one and all ailments causes out of place and excessive usage among people. In India easy purchasibility without any prescription adds to the problem of AMR. The antimicrobials are also being used to promote growth, prevent diseases amongst food-producing animals making it a major contributor to the development of AMR [7].

Even, the environment has a role in the process of AMR. The soil is regarded as a pool of antimicrobial defiant genes and also the water potentially polluted with fecal microbes and organic fertilizers used on plants may spread drug-resistant microbes in the soil. Pharmaceuticals discharges and other chemicals from industries, agricultural facilities, human and animal healthcare facilities, and wastewater treatment plants can also contribute, usually through the discharge in surface waters. Resistant bacteria spread on direct or indirect contact between the human, animal, and environmental reservoirs [8].

Looking at the severity of the problem of AMR an urgent worldwide action plan is needed to control it. The crisis needs to be targeted as One Health perspective because health of human, animals and environment are connected. One Health means joint effort of several sectors in recognizing the interconnection between human, animals, plants, and our environment in the development of AMR [3].

A-One Health approach to combating AMR in human, animal, and environmental sectors requires sustained governmental and private-sector partnership with a commitment to surveillance, research, education, and innovations. Risk management measures that do not involve antimicrobial administration, such as infection control practices in human hospitals and animal care settings, should be identified and evaluated for their role in reducing AMR. Public awareness and education are critical for preserving the effectiveness of existing antimicrobials. Informing the public about appropriate antimicrobials use and providing support for physicians, veterinarians and other healthcare professionals in decision-making are also important. For example, support could take the form of improving the availability of more rapid, sensitive, and specific diagnostic tests for use in clinics and farms, enabling more accurate and prompt treatment decisions.

Keeping into deliberation the magnitude of the burden of AMR and the future dreadful outcomes, the World Health Assembly (WHA) implemented the "Global Action Plan (GAP)" in 2015 [9]. Afterward, other nations developed their own national action plans (NAPs). The "Ministry of Health and Family Welfare" India also prepared its "National Action Plan" in 2017. This five-year NAP (2017–2021) emphasizes the main concerns and implementation approaches for control of AMR control in our country [10]. In 2016, various pharmaceutical, biotechnology, and diagnostics industries signed the "Davos Declaration" on fighting antimicrobial resistance [11].

AMR prevention and control needs to be approached through multi level approach.in which public health experts and community need to participate. key stakeholders need to put collective efforts to control and mitigate the AMR.

10.4 Behavior change and its Theories

Higher education institutes (HEIs) are the centers of innovation and creativity and can have huge influence on the health of communities. These institutes play an role in skill development,education and research. They hold skilled and intellectual expertise and authority, besides enjoying trustworthiness in developing and disseminating knowledge, skills and sustainable innovative solutions to assist behavior change among the target audience. The achievement of the health-promoting movement depends on the ability of educational institutes to incorporate an assurance to health within the policies and procedures.

There is a need to change the key drivers of AMR. The incorrect practice of availability, and use of antimicrobials without prescription needs to be addressed. Increased awareness and sustainable solutions can bring about changes in behavior and are a public health priority against Antimicrobials resistance (AMR) globally. Suitable theory should be incorporated to design and assess interventions for behavior change [12].

A theory helps in the understanding of the probable processes through which the intervention will attain success in behavior modification. Theory-based interventions connect unrelated elements in a logical and well-organized way to capitalize on the utilization of resources [13]. There are numerous behavior change theories and models which discuss the factors influencing behaviors and provide bases for the development of behavioral modification interventions. The health belief model, "Theory of planned behavior," "Diffusion of innovation theory," "Social cognitive theory," "Transtheoretical model," "Social norms theory" are a few of the theories that underpinned the important aspects of behavior change.

Theory of Planned Behavior (TPB) provides a framework to comprehend and foresee behaviors. An important assumption is that behaviors are under voluntary control and are determined by people's intention to carry out that behavior. Attitudes toward the behavior; subjective norms; perceived behavioral control are the three factors that determine people's intention to carry out/ not carry out that behavior.

Intention to change behavior in the context of AMR can be achieved through creating awareness about AMR and may bring positive changes in the attitude of prescribers, users, and society towards antimicrobials usage. Behavioral control regarding antimicrobials prescription by prescribers can be achieved through awareness, research and innovations for, e.g, easy, accessible and rapid diagnostic tools to aid in diagnosis and treatment.

Educational institutes can take many initiatives to reduce unnecessary antibiotics prescription and usage. Educational programs such as campaigns [12], seminars, education using evidence-based medicine and communication skills, educational camps, flyer or leaflets and a combination of any of these strategies can be started and will be very effective to reduce usage of antibiotics. Educational programs have also been found to reduce the prescription of antibiotic.

Other strategies which have been utilized as behavioral change programs are clinical reminders, feedbacks, audits, and financial incentives. The combined intervention methods are more successful in comparison with a single method for patient education [14], delayed prescriptions [15], and decision support system [16]. Multiple interventions have been found to be more effective in reducing the antibiotics prescription when compared with single intervention like creating awareness alone.

"e-Bug teaching pack" is an innovative program implemented in certain European countries for educating school children about the infective microorganisms and their control [17]. Advisory Group on "Antimicrobial Resistance and Healthcare-Associated Infection" (ARHAI) Education sub-Group decided to organise an antimicrobial drive in which posters were prepared to support the people in knowing the ways to handle an illness like common cold.

Physicians are also being motivated to share information with the people regarding the anticipated course of symptoms, self-care, benefits, and ill effects of using antibiotics. The "General Medical Council" in 2009 appealed to all HEIs' Heads to make certain that prevention and control of contagion and prescription of antibiotics should be communicated in the hospital settings to the medical students [18].

10.5 Role of Higher Education Institutes

HEIs can take up the accountability of creating awareness among people with the use of traditional and advanced tools. Various approaches can be used by the HEIs in the direction of the prevention and control of AMR. Approaches towards the advancement of behavioral modifications, development of creative way outs and policies are a few that can aid in attaining the aim. The priority audience for such approaches comprises of the people, health care prescribers, pharmacists, and the policymakers.

Research in the USA has shown that multifaceted campaigns repeated over some time are very successful in creating awareness about prudent antibiotic use. Therefore, a multichannel approach combining mass media, interpersonal channels, and education of young needs to be adopted by HEIs [19].

A national awareness campaign in Belgium and France targeted at antibiotic prescribers stimulated a European Antibiotic Awareness Day on 18th November every year. Important messages, symbols, slogans, and media toolkits, and other campaign material based on scientific evidence were developed. Similar initiatives to develop informative materials and organizations of Awareness campaigns can be initiated by medical and non-medical students of HEIs [20].

In HEIs, lectures, workshops, and seminars on antibiotic usage, along with internet and media campaigns, etc., should be conducted to raise the awareness and modify the behavior of both the medical students and the non-medical students of universities concerning the appropriate use of antibiotics [21]. Interventions and methodologies need to be developed and customized in order to empower the prescribers and the public about the rational and empirical usage of antibiotics. It is critical to develop proper curriculum for training health professionals on microbial virulence, general medicine, antibiotic resistance, and thoughtful recommending of antibiotics [22]. It is also observed that there is irrational use of medication and self-medication among adolescents across the globe. The knowledge of antibiotics is very limited among both medical and non-medical students. So, it is very important to create awareness programs by policymakers for adolescents to upgrade their knowledge on the safe use of antibiotics as well as to prevent antibiotic resistance. These programs can be school and community based [23].

At the school level, the children can be informed about antimicrobials and their ill effects through simple teaching methodologies like lectures and role-plays, animated videos, stories, and games. The AMR and its impact can be integrated into the school curriculum. Various institutes under HEIs have a huge scope in designing and delivery of such information resources. The awareness and perception about antibiotic usage and resistance amongst school teachers is not very promising. School education programs and public education are imperative to modify the behavior related to antibiotic usage amongst all the stakeholders. It will aid in the control of antimicrobials resistance and thus help in the preservation of antibiotics for coming generations [24].

At the public level, HEIs can design mobile apps, key messages, slogans for disseminating information in an easy, accessible, and affordable way. Media schools can create short movies and advertisements/teasers for creating awareness at the local, national and global levels. Inappropriate demands, usage, and ineffective usage can be brought down through these awareness strategies. The role of the public in the development of AMR can be made aware to the general public. The importance of hand hygiene and basic sanitation should be the key messages in the prevention of AMR.

It has been reported that one of the major causes of misuse of antimicrobials is a lack of sufficient knowledge of antimicrobial prescribing [25]. Various studies indicate that knowledge and training received by medical students of HEIs about antibiotic use and bacterial resistance are not sufficient [26]. Behavior change can also be brought about by changing the prescribing etiquette followed by medical professionals. The academic researchers in HEIs can identify the drivers of prescribing behavior and gaps in action taken to bring about changes in antimicrobial prescribing [27]. The medical students are future prescribers and it is important to understand their perceptions, attitudes, and knowledge about antimicrobial prescribing practices. By educating these future doctors about appropriate use of antibiotics and antibiotic stewardship the problem of antimicrobial resistance can be addressed to a large extent [28].

Educating health professionals about antibiotic stewardship can transform into optimal antibiotic prescribing. This education and training can also be extended to medical students. Both prescriber's knowledge and behavior can be changed by using passive and active techniques of education. Passive techniques are moderately effective for increasing prescriber knowledge, whereas active techniques are more effective for changing the prescribing behavior. These education techniques can very effectively increase knowledge and the use of antibiotics. Future doctors need to be given adequate training about proper prescribing, dispensing, and usage of antibiotics [29].

HEIs with teaching hospitals can contribute to the development of training curriculum for medical professionals and can initiate the "Train the Trainer" program and training programs as nodal centers. The training programs should not be limited to medical concepts of AMR but also focus on interpersonal communication to counsel patients on adverse effects of antimicrobials as irrational pressure from patients to prescribe/dispense antimicrobials also serves as a reason to prescription. They should be made acquainted with best infection prevention and control (IPC) practices and motivated to comply with such practices for the prevention of healthcare-associated infections and the emergence of AMR. Commitment posters for stewardship and IPC can be designed and displayed in various health care settings. Handy and easily accessible information on the indication, selection of correct antibiotic, dosage, and duration can be developed and made available to prescribers so that misuse of antibiotics can be prevented.

Interpersonal training in communication for pharmacists can be imparted by HEIs to get better at handling patients' demands for unnecessary antimicrobials. Behavioral modification campaigns should focus on motivating pharmacists to sell antimicrobial medications based on medical prescriptions only and to provide adherence counseling to the patients at the right time. Even the pharmaceutical companies need to make aware of their contribution in aggravating the AMR by the discharge of pharmaceutical effluents in rivers and poisoning the soil and leading to the emergence of resistant microbial strains.

Students of HEIs can help parents understand the importance of prudent use of antibiotics as their knowledge of antibiotics is poor. By investing in their education, informed decision-making by parents can be encouraged [30]. The training resources and other tools on AMR can be made available via dedicated AMR web pages. A booklet available at the e-Bug website named "When Should I Worry?" developed by Cardiff University provides information about common infections in children. The use of this booklet resulted in tremendous reduction in antibiotic prescribtion. Taking motivation from, other HEIs can develop such resources and share these with the public to improve their knowledge about the misuse and overuse of antibiotics. Such resources will help to improve public knowledge about antibiotic resistance and also increase their faith in self-care for the majority of self-limiting infections, without consulting their doctor.

AMR is a problem which involves various disciplines, and requires a One Health comprehensive approach from all stakeholders for control [31]. The students of HEIs can also be involved in designing and implementing AMR awareness campaigns for the animal and food production sectors. The students of HEIs can engage local communities through a variety of activities and can change their knowledge and behavior towards the use of antimicrobials. They should focus on clear and concise messages while ensuring appropriate message delivery to the audience. To create awareness among the public about the misuse and overuse of antibiotics, HEIs can develop a comprehensive communication strategy with the help of students. At the same time, it should not be assumed that a higher level of awareness will always bring change in behavior as it needs a dynamic and long process that rests on four pillars: adequate awareness, robust regulatory environment, emotional or material incentives, and an enabling social structure. All these pillars need to be addressed satisfactorily [32].

HEIs can collaborate at local, national, and global levels for the development of technologies and solutions against AMR. "PREVENT IT" is a global project co-funded by the Erasmus+ program of the European Commission. The project is implemented by a consortium comprising 5 Indian universities, 4 European universities, and 2 non-governmental organizations having super-expertise in different health sciences fields and societal contexts. The project is aiming at a future where every onel is aware of antibiotic resistance (ABR) and and has correct knowledge on how to address the issue of ABR. It aims to promote qualitative teaching activities in academia as well as in society. Modules on ABR have been included in the syllabus of under graduate students to provide them required knowledge on causes and prevention and control of ABR. The experts have recognized that policymakers frequently encounter issues like incorrect information, limited access to technical expertise to weigh the impact of AMR and a lack of information on available options, all of which hamper the making of critical policies aimed at supporting investments and actions to fight AMR. Therefore, HEIs can aid in developing evidence-based and multiple data-driven policy briefs to acquaint the policymakers about current issues of AMR and the way forwards to combat it.

10.6 Conclusion

Antimicrobial resistance is a complex and multifactorial global health issue. A multitude of factors at public, health care, environmental, animal production, and governmental levels are responsible for its emergence, acceleration and spread and thus have an enormous scope of improvement. There is an emergent need to target AMR throughout its trajectory starting from its mechanism of origin to means of its prevention and control. A multidisciplinary team consisting of scientists, health care providers, sociologists, engineers, communication specialists, and policymakers are required to conquer AMR. Higher education institutes having experts and requisite resources can serve as an imperative stakeholder in the fight against it through research, innovation, Information Education Communication (IEC), and acceptable, accessible and affordable, technological solutions for behavioral change under One Health Perspective. HEIs need to understand their distinctive position to influence society, therefore, must accept their responsibility towards the attainment of SDG-3 through prevention and control of AMR. Policymakers at local, national and international levels must recognize the potential of HEIs to assimilate multidisciplinary pieces of evidence and solutions against AMR.

Bibliography

[1] SJ Bright-Ponte, BK Walters, H Tate, LM Durso, JM Whichard, KE Bjork, CB Shivley, AL Beaudoin, KL Cook, EL Thacker, et al. One health and antimicrobial resistance, a united states perspective. *Revue scientifique et technique (International Office of Epizootics)*, 38(1):173–184, 2019.

[2] Francesca Prestinaci, Patrizio Pezzotti, and Annalisa Pantosti. Antimicrobial resistance: a global multifaceted phenomenon. *Pathogens and global health*, 109(7):309–318, 2015.

[3] CDC. CDC funds 34 innovative projects to combat antibiotic resistance — CDC Online Newsroom — CDC, 2016.

[4] Björn Bengtsson and Christina Greko. Antibiotic resistance—consequences for animal health, welfare, and food production. *Upsala journal of medical sciences*, 119(2):96–102, 2014.

[5] Paul Flowers. Antimicrobial resistance: A biopsychosocial problem requiring innovative interdisciplinary and imaginative interventions. *Journal of Infection Prevention*, 19(4):195–199, 2018.

[6] Robert A Weinstein. Controlling antimicrobial resistance in hospitals: infection control and use of antibiotics. *Emerging infectious diseases*, 7(2):188, 2001.

[7] Bonnie M Marshall and Stuart B Levy. Food animals and antimicrobials: impacts on human health. *Clinical microbiology reviews*, 24(4):718–733, 2011.

[8] Patrick Boerlin and David G White. Antimicrobial resistance

and its epidemiology. *Antimicrobial therapy in veterinary medicine*, pages 21–40, 2013.

[9] WHO. Antimicrobial resistance, 2019.

[10] World Health Organization. Antimicrobial Stewardship Programme in Health-Care Facilities in Low and Middle Income Countries, 2019.

[11] Joan A Casey, Brent F Kim, Jesper Larsen, Lance B Price, and Keeve E Nachman. Industrial food animal production and community health. *Current environmental health reports*, 2(3):259–271, 2015.

[12] Meenakshi Sood, Vinay Kumari, and Bunty Sharma. Behavior changes: Role of higher education institutes in combating amr. *SPAST Abstracts*, 1(01), 2021.

[13] Julian Davies and Dorothy Davies. Origins and evolution of antibiotic resistance. *Microbiology and molecular biology reviews*, 74(3):417–433, 2010.

[14] The Federal Council. Strategy on Antibiotic Resistance Switzerland. page 74, 2015.

[15] Richard H Gustafson and RE Bowen. Antibiotic use in animal agriculture. *Journal of applied microbiology*, 83(5):531–541, 1997.

[16] National Water Agency (PUB) Agri-Food Veterinary Authority of Singapore (AVA), Ministry of Health (MOH), National Environment Agency (NEA). Singapore: National strategic action plan on antimicrobial resistance, 2017.

[17] Donna M Lecky and Cliodna AM McNulty. Current initiatives to improve prudent antibiotic use amongst school-aged children. *Journal of Antimicrobial Chemotherapy*, 68(11):2428–2430, 2013.

[18] Cliodna AM McNulty, Barry D Cookson, and Michael AO Lewis. Education of healthcare professionals and the public.

Journal of antimicrobial chemotherapy, 67(suppl_1):i11–i18, 2012.

[19] Mary Jo Trepka, Edward A Belongia, Po-Huang Chyou, Jeffrey P Davis, and Benjamin Schwartz. The effect of a community intervention trial on parental knowledge and awareness of antibiotic resistance and appropriate antibiotic use in children. *Pediatrics*, 107(1):e6–e6, 2001.

[20] Sarah Earnshaw, DL Monnet, B Duncan, J O'Toole, K Ekdahl, H Goossens, European Antibiotic Awareness Day Technical Advisory Committee, European Antibiotic Awareness Day Collaborative Group, et al. European antibiotic awareness day, 2008–the first europe-wide public information campaign on prudent antibiotic use: methods and survey of activities in participating countries. *Eurosurveillance*, 14(30):19280, 2009.

[21] Poonam Shah, Rajeev Shrestha, Zongfu Mao, Yilin Chen, Yan Chen, Pramesh Koju, Xinliang Liu, and Hao Li. Knowledge, attitude, and practice associated with antibiotic use among university students: a survey in nepal. *International Journal of Environmental Research and Public Health*, 16(20):3996, 2019.

[22] Chang-Ro Lee, Jung Hun Lee, Lin-Woo Kang, Byeong Chul Jeong, and Sang Hee Lee. Educational effectiveness, target, and content for prudent antibiotic use. *BioMed research international*, 2015, 2015.

[23] Satish Kumar Deo, Sita Rijal, Sita Devi Kunwar, Anuja Dahal, and Sujaya Gupta. Knowledge of use of antibiotic, its resistance and consequences among students in private schools. *Cough*, 123:26–2, 2018.

[24] Anita Kotwani, Chand Wattal, PC Joshi, and Kathleen Holloway. Knowledge and perceptions on antibiotic use and resistance among high school students and teachers in new delhi, india: A qualitative study. *Indian Journal of Pharmacology*, 48(4):365, 2016.

[25] Celine Pulcini and Inge C Gyssens. How to educate prescribers in antimicrobial stewardship practices. *Virulence*, 4(2):192–202, 2013.

[26] Luis Felipe Higuita-Gutiérrez, Valentina Molina-Garcia, Jenifer Acevedo Guiral, Liceth Gómez Cadena, Gustavo Eduardo Roncancio Villamil, and Judy Natalia Jiménez Quiceno. Knowledge regarding antibiotic use among students of three medical schools in medellin, colombia: a cross-sectional study. *BMC Medical Education*, 20(1):1–9, 2020.

[27] Esmita Charani, Enrique Castro-Sánchez, and Alison Holmes. The role of behavior change in antimicrobial stewardship. *Infectious Disease Clinics*, 28(2):169–175, 2014.

[28] Lilian M Abbo, Sara E Cosgrove, Paul S Pottinger, Margaret Pereyra, Ronda Sinkowitz-Cochran, Arjun Srinivasan, David J Webb, and Thomas M Hooton. Medical students' perceptions and knowledge about antimicrobial stewardship: how are we educating our future prescribers? *Clinical infectious diseases*, 57(5):631–638, 2013.

[29] Ruchi Shrestha. Knowledge, attitude and practice on antibiotics use and its resistance among medical students in a tertiary care hospital. *JNMA: Journal of the Nepal Medical Association*, 57(216):74, 2019.

[30] D Donà, E Barbieri, M Daverio, R Lundin, C Giaquinto, T Zaoutis, and M Sharland. Implementation and impact of pediatric antimicrobial stewardship programs: a systematic scoping review. *Antimicrobial Resistance & Infection Control*, 9(1):1–12, 2020.

[31] Roksana Hoque, Syed Masud Ahmed, Nahitun Naher, Mohammad Aminul Islam, Emily K Rousham, Bushra Zarin Islam, and Shaikh Hassan. Tackling antimicrobial resistance in bangladesh: A scoping review of policy and practice in human, animal and environment sectors. *PloS one*, 15(1):e0227947, 2020.

[32] Philip Mathew, Satya Sivaraman, and Sujith Chandy. Communication strategies for improving public awareness on appropriate antibiotic use: Bridging a vital gap for action on antibiotic resistance. *Journal of family medicine and primary care*, 8(6):1867, 2019.

11

Behavioural Change: Role of NGOs to Combat AMR

Dona Boban

Amrita Institute of Medical Science (AIMS), Kochi, India

Kiranjeet Kaur

Chitkara School of Health Sciences, Chitkara University, Punjab, India

Sanjeev K. Singh

Amrita Institute of Medical Science (AIMS), Kochi, India

CONTENTS

11.1 Introduction

Antimicrobial resistance (AMR) is a phenomenon in which the microbes change over time and do not respond to the medicines which were once effective against them. It is a natural phenomenon accelerated by many anthropogenic factors. Human behaviours relating to antimicrobial use is the prime driver of AMR globally. It seems that the use of antimicrobials in humans, animals, plants or the environment is majorly dependent on humans. The injudicious antimicrobial use is driven by over-the-counter use of antimicrobials, non-prescription use, incomplete courses of medications, sharing of medications to name a few. Additionally, the effluents containing antimicrobials from the pharmaceutical industries are driving forward the AMR in the environment. The issue of AMR seems more complex when one understands that the drug development pipeline is dry and not enough research and development is happening for the discovery of newer antibiotics with potent mechanisms. Lack of awareness on resistance among healthcare workers, policy makers, public health professionals and the general public escalates the issue.

Antimicrobials are considered societal drugs, which means use of antimicrobial by any person can have an impact on another person, animal or in the environment. Hence, while trying to address the issue of AMR, the solutions need to be of a "shared value" approach. In other words, the solutions should focus on maintaining the benefit to patients without any compromise of societal health. This requires a multi-level, multi stakeholder approach including the state, private sector and the non-governmental organisations (NGOs).

AMR is not new. It was rightly anticipated by Alexander Fleming at the time of the discovery of the first antibiotic, back in 1928. The danger of AMR was overshadowed by the use of a variety of antimicrobials leading to decrease in morbidity and mortality. The world started to recognise AMR when the antimicrobials started to fail. Looking at the solutions in place by Global and National action plan, the improved awareness through education and training seems one of the best and easy solutions targeting the problem at grass root level [1]. The challenge is to outreach to the health care workers, academicians, students and general public. It seems that the latter one can be targeted easily by the support of Non-Governmental Organisations (NGOs).

The NGOs can play a key role in letting the world know the danger of AMR. They play a prominent role in improving awareness and perceptions to manage AMR. NGOs are the voice of the general public to the policymakers. Their role spans across the domains of policy, advocacy, innovation and research, surveillance, and funding.

11.2 Human behaviour and AMR

Human behaviour is the "*range of behaviours performed by humans, influenced by many factors like culture, attitudes, emotions, values, ethics, authority, rapport, persuasion, coercion, and/or genetics*" [2]. There is a human behaviour component on all aspects of AMR because we handle these miracle drugs [3]. It has been observed that to bring a sustainable change towards the rational use of antibiotics – change in knowledge, attitude and practice are the major drivers [4]. Understanding human behavior offers opportunities to develop an effective communication strategy. AMR will be a persisting threat as long as the human behaviour and practices do not change.

The various situations wherein the human behaviour is involved in relation to antimicrobials and AMR includes prescribing, dispensing, consuming and discarding antimicrobials, non human use of antimicrobials, environmental spread of antimicrobial residues and resistant bacteria and genes, infection prevention and control and increased travel.

The decision to prescribe or consume an antimicrobial depends on various macro level and micro level factors. Various determinants of antimicrobial use at macro level includes drug policy, legislation, clinical guidelines, education, resources, culture, pharmaceutical companies, availability and accessibility of antimicrobials and the disease and resistance patterns [2]. Additionally, there are various micro level factors influencing the decision making process. These include knowledge and experiences, perceived risks/benefits, social influence, practice related factors, personal advocacy, self efficacy to list a few [2].

The common behaviours of prescribers and dispensers which drive forward AMR include providing antibiotic to the customer without prescription, providing low quality/ counterfeit/ fake drugs, prescribing antibiotics for cases that are not indicated, etc. From the side of consumers, we can identify various self medication practices including inadequate dose, non completion of treatment, sharing of antibiotics and improper disposal of antibiotics. The reason why people resort to such practices are the notion that the people by themselves can take the drug of choice for their illness, the inbuilt belief that antibiotic can cure any disease, easy over the counter access to antimicrobials, or fear of loss of daily wages due to illness. The human behaviours driving AMR is not restricted to human use of antimicrobials. To meet the increasing protein demand due to population growth, farmers resort to intensive farming, where large number of animals are reared in closed, confined spaces with compromised hygiene and sanitation. To prevent the spread of any infectious disease within the flock, farmers provide low doses of antimicrobials along with the feed to the animals, creating a perfect niche for the growth and spread of resistant microbes and genes. It is interesting to know that the pastoralist tribes in Ethiopia uses small amounts of antibiotics in the milk to be sold to cover up any potential infections due to unsterilised milk vessels [5].

In rural areas, households share areas with livestock for living. This is the perfect opportunity for transmission of both resistant bacteria and their genes via direct contact with animal and its feces. Similarly, there are many behaviours relating to slaughtering of animals like- Handwashing using soap after slaughtering poultry

1 in 18 cases in households. The animal carcass or waste are disposed on open areas. These are eaten by dogs, birds and poultry. In various small-scale animal food processing facilities, the measures of biosecurity are often meagre or absent. For example, there is no soap or running water during slaughtering, no regulations are followed during poultry slaughtering, liquid waste from markets are disposed into drains directly. Additionally, many cultures across the world eat raw or undercooked meat which is a common means of transmission of ABR bacteria and gene to humans.

Similarly, it is a common in rural areas to see humans and animals using water from the same river for washing, bathing and drinking. This results in easy transmission of AMR bacteria and genes. The seasonal pilgrimage to sacred rivers are also increasing the antimicrobial resistant gene contamination of rural river waters. Regarding human interaction with surrounding environment, we can see that the environment is getting contaminated with AMR genes and residual antibiotics due to human activities in various sites including lapses in implementing the regulations and not checking the maximum residual limits in the effluents.

The spread of resistant microbes and genes across the globe has made it difficult for countries with strong disease control programs to treat an infection. Travelers to regions with high AMR can be exposed to multi-drug resistant organisms and return to their countries colonized and be vectors. When the traveller is avoiding the vaccination, not being compliant with hygiene practices and is self medicating with antibiotics, he/she are putting themselves at risk to contract an AMR infection. In such situations, they are at high risk of contracting or spreading infections including the resistant ones.

Considering the above mentioned human behaviours, the question arises whether it is possible to change these behaviours? The answer is a definite yes. These can be changed at various levels [2] as shown in the image (Figure 11.1.).

At individual level, a change in knowledge, attitudes, beliefs and personality can lead to a change of behaviour. The social

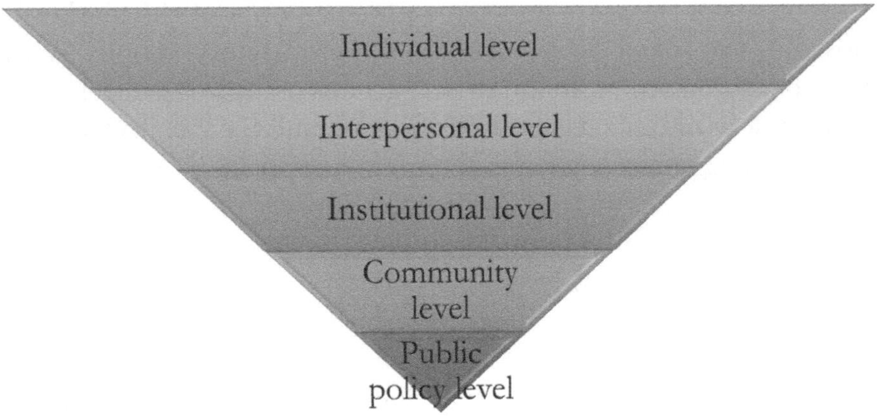

FIGURE 11.1
Social Ecological Model of Behaviour Change

identity, social support and societal roles influence the behaviour change at interpersonal level. Development and implementation of rules, guidelines, regulations plays active role at the institutional level. Social networks, social norms and beliefs, influence of NGOs on the communities influence the changes at community level. At public policy level, regulations and laws are the key to influence the behaviour. The NGOs come into action at each levels through improving the awareness on AMR, advocacy activities, campaigning, funding to list a few. One way of achieving the same is through the development of the Information, Education and Communication (IECs) for different audiences depending on their need. The detailed discussion on the role of NGOs follows.

11.3 Non- Governmental Organisations (NGO) or (CSO)

Civil Society Organisation (CSO) NGO or Civil Society Organisation (CSO) is an organisation that functions independently of any government. The World Health Organisation (WHO) defines CSO as "the social arena that exists between the state and the individual or household. Civil society lacks the coercive or regulatory power of the state and the economic power of the market but provides the social power or influence of ordinary people. Within this social domain, individuals and groups organize themselves into civil society organizations (CSOs) to pursue their collective interests and engage in activities of public importance" (WHO 2001a). Most of the NGOs are not for profit, voluntary organisations. They are organised at community or grassroots level, regional level, national or international level and work towards a social or political cause. The World Bank identifies two different types of NGOs [6]:

1. Operational NGOs – that focus on design and implementation of development projects. They try to achieve small scale changes directly through projects.

2. Advocacy NGOs - that defend or promote a specific cause and try to influence public policy. They try to achieve large scale changes by influencing the political decisions [7]

Some NGOs might fall under both the categories like those working in the health sector. NGOs play a critical role in promoting social changes in the community. They are the voice of the people. They put forth the needs of people in front of the decision makers. They also play an imperative role in building healthy communities and improving citizen participation. They address a multitude of issues, especially in health.

11.3.0.1 Role of NGOs in health sector

NGOs were involved with public health for a long time. Records show that they were active in cleaning up American cities in the 1800s. The true recognition for NGOs came during the 1978 Alma Ata Declaration. It recognised people's participation in the healthcare system as central to primary health care and recognised the role of NGOs in securing health gains. The NGOs play a critical role in health services through service provision, facilitating community interactions, provision of resources and supporting the health worker. Their role in health promotion consists of collecting and disseminating health information, conducting research studies and utilising the results to decide further actions, building informed public choice and helping to change the social attitudes surrounding a health issue. They also play an important role in resource mobilisation and allocation through financing the services, mobilising and organising community co-financing schemes and promoting equity in resource allocation. They are the voice of the marginalised community. They represent patient rights in issues relating to quality of care. They are also involved in monitoring the quality of healthcare delivery. Under the realm of policy making, NGOs plays a significant role in representing the public interest of the people, promoting equity and pro-poor policies, building consensus, disseminating policy information and raising public support for policies [8]. NGOs are actively involved in various public health issues, AMR being one among them.

11.3.0.2 Role of NGOs in AMR prevention and control activities

AMR is a One Health issue, and this is exactly why NGOs have an important role in the prevention and control activities pertaining to AMR. To address the issue of AMR, a global collaboration across various sectors is essential. NGOs are already working in various sectors, and they are the best to turn to for insights on how to collaborate better under One Health approach. NGOs have specialised knowledge and capacity to act swiftly across various levels, offering a platform for strong public engagement. The role of NGOs in curbing AMR is not limited to advocacy. They are involved at multiple levels viz.

- regulating the marketing practices of pharmaceutical industries
- improving access to antimicrobials, vaccines and diagnostics
- campaigning to ban non therapeutic use of antimicrobials in animals and plants
- research activities
- influencing policy decisions

NGOs can reach out to the communities and empower them with the right messages and implement practices on rational use of antimicrobials. They are knowledgeable about the local context. Hence, they could very well tailor the awareness messages and good practices on judicious use of antimicrobials based on the culture and the local context of the communities. Additionally, they can be of help to plan and implement operational research projects in

AMR to find solutions at the grassroot level. NGOs can sustain and amplify communication to create awareness on AMR. NGOs have a significant role in engaging with policymakers and politicians through strong advocacy statements. Additionally, they can mobilise key stakeholders and push for collective action. To rightly address the issue of AMR, communities need to be involved and their needs and realities must be studied. It becomes very important to identify the drivers of AMR and to plan interventions according to the target audience. The NGOs can provide insights on the needs of the people and the much needed action. The Bangkok declaration has rightly mentioned the same. It has been stated that "civil society has a role to play to turn the rising tide of drug-resistance infections by catalysing change, mobilizing support and resources for implementation and ensuring [the] accountability of our food system."

11.3.0.3 International NGOs and their role in curbing AMR

NGOs are playing a key role in promoting right knowledge on judicious antimicrobial use to curb the issue of AMR. They were active from the 1980s onwards working towards the same cause. The organisations like Health Action International (HAI), Oxfam, International Organisation of Consumer Unions, were working on ending the promotional and marketing activities of pharmaceutical industries wherein the antibiotics were sold directly to consumers. Various professional societies like Infectious Diseases Society of America, American Academy of Paediatrics developed and continuously updates the antibiotic treatment guidelines for the health-care workers, to save the efficacy of these miracle drugs. Consumer International is an organisation that focuses their activities on reducing non-therapeutic use of antimicrobials in animal husbandry.

The WHO recognised that AMR is a global public health issue in the year 1998 and adopted a resolution to take action on judicious use of antimicrobials, research and development. WHO came up with a Global Strategy for Containment of Antimicrobial Resistance in 2001 [9]. Action on Antibiotic Resistance, also named as ReAct, is a global network that work hand in hand with WHO and published "The evolving threat of antimicrobial resistance:

options for actions" [10]. Following this, the momentum picked up and more organisations got involved in activities to prevent and control AMR.

In the 2013 Global risk report by the World Economic Forum, antibiotic resistance was highlighted [11]. During this time, the NGOs like Médecins sans Frontières (MSF) and Health Action International (HAI) worked on fair access to drugs and vaccines. Organisations like Alliance to Save our Antibiotics and Food Animal Concerns Trust focussed on decreasing non-therapeutic antimicrobial use in animals through a regulatory approach. In view of the increased recognition that AMR is a one health issue, ReAct proposed intersectoral coalition of civil society groups to tackle AMR and developed Declaration of Shared Principles. This includes non-human use of antibiotics, innovation, sustainability, access to antibiotics, use of antibiotics by humans, and systems thinking. The Declaration on Antibiotic Resistance (ABR) was built around these 5 principles. The finalised declaration and unveiling of the Antibiotic Resistance Coalition (ARC) occurred during the 2014 World Health Assembly wherein the members also adopted a resolution to develop a Global Action Plan to prevent and control AMR. Twenty organisations supported the declaration including Public Citizen, the Centre for Science in the Public Interest USA, Centre for Science and the Environment India, HAI, Third World Network, the People's Health Movement, and Universities Allied for Essential Medicines. Over time, ARC grew, and more organisations became part of the coalition like Natural Resources Defence Council, the Ecumenical Pharmaceutical Network (EPN), the American Medical Student Association, MedAct and the US Public Interest Research Group (PIRG). ARC continues to be a stage for the members to converse about key political issues on AMR, share experiences by expertise across various sectors, and organise communal responses across various forums. The annual WHO-NGO dialogue organised by ARC, right before the World Health Assembly acts as the point of contact between NGOs and the key AMR leadership at WHO to discuss the important issues on AMR policy processes and can influence the policy decisions. ARC also advocated inclusion of the abovementioned principles in the Political Declaration of the UN meeting on AMR. Let's discuss further on how NGOs contributed to these five pillars.

11.3.0.4 NGO role in human and non-human use of antibiotics

The NGOs Health Care Without Harm and PIRG engaged health-care professionals on judicious use of antibiotics in human and non-human scenarios. Health Care Without Harm developed guidelines for the procurement meat and seafood products raised without antibiotics by the hospitals. They established Healthy food in Health Care programs to increase the market demand for antibiotic free animal products. They also formed Clinician Champions in Comprehensive Antibiotic Stewardship (CCCAS) Collaborative to improve the knowledge on AMR as one health concept among health-care workers. This also acts as a platform to promote policies on judicious use of antibiotics across various sectors. PIRG mobilised health care workers and public health professionals to sensitise restaurant chains to source antibiotic free animal products.

To curb the non-therapeutic antibiotic use in food animals, the NGOs like Consumers Union, PIRG, the Centre for Food Safety, Friends of the Earth, the Natural Resources Defence Council (NRDC) and the Food Animals Concerns Trust demanded the famous restaurant chains like Subway, McDonald's, Taco Bell, etc, to source the animal products that are antibiotic free. The restaurants came under consumer pressure and began to respond positively to the demands raised by the NGOs. The NGO EPN is involved in education campaigns, and research on knowledge on antibiotic use and resistance [12].

11.3.0.5 NGO role in innovation

NGOs are playing a commendable role in documenting the lack of new antibiotics under research and development. The ReAct group along with European Medicines Agency and the European Centre for Disease Prevention and Control found out that there is no new antibiotic with an innovative mechanism against gram negative pathogens. This evidence led to the eventual establishment of the Transatlantic Task Force on Antimicrobial Resistance (TATFAR). The civil society groups played the prime role in bringing forward the fact that the pharmaceutical industries were slow in acknowledging- the fact that there are no new antibiotics in the pipeline. The NGOs like ReAct brought together stakeholders from public and private sectors to discuss research and development of antibiotics. The stakeholders included TB Alliance, India's Open-Source Drug Discovery Initiative, and the Drugs for Neglected Diseases Initiative (DNDi). This was followed by the launch of Global Antibiotic Research and Development Partnership (GARDP). Antibiotic Action affiliated to British Society of Antimicrobial Chemotherapy, works to raise awareness of the need for new antibiotics and promotes the research and development of the same [13].

11.3.0.6 NGO role in AMR in environment

The NGO Changing Markets Foundations exposed the toxic levels of antibiotic effluents pumped out into the environment from various Indian and Chinese drug manufacturing companies. Other civil society groups also came forward to discuss and find a sustainable solution to this issue. The continuous advocacy by the NGOs led to the 2016 Davos "Declaration by the Pharmaceutical, Biotechnology and Diagnostics Industries on Combating Antimicrobial Resistance" that mentions measures to curb the environment pollution by antibiotics. Later that year, the "Industry Roadmap for Progress on Combating Antimicrobial Resistance" was signed by some companies which called for several measures to stop the environment pollution during antibiotic production. The Health Care without Harm NGO examined the waste management practices in the hospitals in the USA and suggested a no burn approach as a safer alternative. The Centre for Science and Environment in India tried to integrate the one health aspect into the national action plans to curb AMR in developing countries.

11.3.0.7 NGO role in surveillance

NGOs play a crucial role in regions where no surveillance data on AMR are available, especially in humanitarian emergency situations. For example, MSF documented the prevalence of resistance bacteria in many research studies. MSF shares all data on AMR with the WHONET. They developed the Mini Lab project wherein they established a primary microbiology culture laboratory in resource limited setups. They also conduct surveillance of drug resistant tuberculosis in areas where they work. This catalysed the setting up of a surveillance system by the national government. The EPN is supporting the surveillance of antibiotic use across 15 countries [12].

11.3.0.8 NGO role in diagnostics and vaccines

NGOs like MSF and ReAct along with Foundation for Innovative New Diagnostics (FIND) and WHO brought together experts to discuss the development of diagnostics to distinguish bacterial infections in acute fever cases. MSF did a global campaign to bring down the price of pneumococcal vaccine, which could in turn increase the accessibility to the vaccines. This could in turn significantly avert antibiotic therapy.

11.3.0.9 NGO role in monitoring and accountability

NGOs are actively involved in monitoring and evaluating the implementation of various health initiatives. Many of them have already started to evaluate the implementation of national action plans on AMR. They are also involved in analysing the progress of the international organisations and industries commitments to address AMR. The ARC helps strengthen transnational advocacy, pressurising government, organisation, and industries to keep AMR high in their priorities. Further, ARC has engagements with various stakeholders like healthcare workers, farmers, consumers etc. This helps them to monitor the developments, implementation of commitments and build campaigns to demand more action.

11.4 Other NGOs actively involved in AMR prevention and control activities

Other NGOs actively involved in AMR prevention and control activities are as follows:

11.4.0.1 The Pew Charitable trusts

It is a global non-governmental organisation that address AMR by establishing and monitoring data driven targets on hospital antibiotic use in collaboration with CDC, developing sustainable antibiotic stewardship practices in food animal industry and providing an open access data sharing and analysis tool named SPARK to drive forward research and development of new antibiotics [14].

11.4.0.2 Centre for a Livable Future (CLF)

CLF is affiliated to Johns Hopkins Bloomberg School of Public Health, working with students, teachers, researchers, policymakers, advocacy groups, to build a healthy, equitable and resilient food system. They revealed that the antibiotic use for promoting growth in poultry is not cost effective. They displayed evidences about the presence of antibiotic residues in the feed for animals and poultry [15, 16]. They also identified that the communities living near the poultry farms are at higher risk to acquire Methicillin Resistant Staphylococcal infection or MRSA [17]. More details about the work of this NGO and their success stories can be accessed through https://clf.jhsph.edu/.

11.4.0.3 Alliance for Prudent Use of Antibiotics (APUA)

APUA is affiliated under the International Society of Antimicrobial Chemotherapy. They provide technical expertise to policy makers, organisations and other stakeholders to improve the decision, supply and use of antimicrobials. They also conduct research and provide technical guidance on antimicrobial treatment and resistance from grassroot to global levels. To know more about their activities, please access their official website through https://apua.org/.

11.4.0.4 World Alliance Against Antimicrobial Resistance (WAAAR)

WAAAR successfully advocated on AMR to the policymakers and legislators in France. The existing programmes on AMR surveillance was reinforced by WAAAR with the help of European Centre for Disease Prevention and Control (ECDC). They established the newsletter "AMR Times" and the yearly publication "AMR

Control". They are also actively involved in many scientific and research programmes. They are pioneers in developing communication programme targeting various stakeholders like politicians, legislators, senate, mayors and other influential people from various regions [13].

11.4.0.5 The Centre for Disease Dynamics, Economics and Policy (CDDEP)

CDDEP is a research and advocacy organisation in the public health arena, with AMR being one among them. They developed the Global Antibiotic Resistance Partnership (GARP), which was instrumental for situational analyses on antibiotic use and resistance, and developed recommendations based on the same across 8 countries [12]. More information can be gathered from the official CDDEP website at https://cddep.org/research-area/antibiotic-resistance/

11.5 Conclusion

A shift in human behaviour is one of the key factors that can bring a big change in society, especially the general public. Specially, if the information is shared in local, regional language with valid data/numbers it can have a greater impact. This role can be played significantly well by the non-governmental organisations (NGOS) that are in constant touch with the general public. Apart from creating awareness, NGOs now actively get involved in innovation, surveillance, monitoring in collaboration with other organisations. Such initiatives seem much more fruitful as they offer opportunities for improving people outreach, information exchanges and increasing resource outputs.

Bibliography

[1] Government of India. National action plan antimicrobial resistance (nap-amr) 2017–2021. 2017.

[2] Cecilia Stålsby Lundborg and Ashok J Tamhankar. Understanding and changing human behaviour—antibiotic mainstreaming as an approach to facilitate modification of provider and consumer behaviour. *Upsala journal of medical sciences*, 119(2):125–133, 2014.

[3] Joseph O Othieno, Obadiah Njagi, and Allan Azegele. Opportunities and challenges in antimicrobial resistance behavior change communication. *One Health*, 11:100171, 2020.

[4] Giacomo Scaioli, Maria R Gualano, Renata Gili, Simona Masucci, Fabrizio Bert, and Roberta Siliquini. Antibiotic use: a cross-sectional survey assessing the knowledge, attitudes and practices amongst students of a school of medicine in italy. *PloS one*, 10(4):e0122476, 2015.

[5] Kebede Amenu, Barbara Wieland, Barbara Szonyi, and Delia Grace. Milk handling practices and consumption behavior among borana pastoralists in southern ethiopia. *Journal of Health, Population and Nutrition*, 38(1):1–12, 2019.

[6] Jean Folger and M James. What is an ngo (non-governmental organization). investopedia, 2021.

[7] Peter Willetts. What is a non-governmental organization, 2002.

[8] FAOOIE WHO. The fao-oie-who collaboration sharing responsibilities and coordinating global activities to address

health risks at the animal-human-ecosystems interfaces. In *International Ministerial Conference on Animal and Pandemic Influenza. Hanoi*, 2010.

[9] World Health Organization et al. Who global strategy for containment of antimicrobial resistance. Technical report, World Health Organization, 2001.

[10] World Health Organization et al. *The evolving threat of antimicrobial resistance: options for action*. World Health Organization, 2012.

[11] Mee-Ngan F Yap. The double life of antibiotics. *Missouri medicine*, 110(4):320, 2013.

[12] EA Ashley, J Recht, A Chua, D Dance, M Dhorda, N Thomas, et al. Antimicrobial resistance in low and middle income countries: an analysis of surveillance networks, 2017.

[13] Jean Carlet, Celine Pulcini, and Laura JV Piddock. Antibiotic resistance: a geopolitical issue. *Clinical Microbiology and Infection*, 20(10):949–953, 2014.

[14] Pew. Organizations across the world commit to fight antimicrobial resistance, 2018.

[15] Patrick A Baron, David C Love, and Keeve E Nachman. Pharmaceuticals and personal care products in chicken meat and other food animal products: a market-basket pilot study. *Science of the total environment*, 490:296–300, 2014.

[16] Amir Sapkota, Amy R Sapkota, Margaret Kucharski, Janelle Burke, Shawn McKenzie, Polly Walker, and Robert Lawrence. Aquaculture practices and potential human health risks: current knowledge and future priorities. *Environment international*, 34(8):1215–1226, 2008.

[17] Joan A Casey, Brent F Kim, Jesper Larsen, Lance B Price, and Keeve E Nachman. Industrial food animal production and community health. *Current environmental health reports*, 2(3):259–271, 2015.

12

Measures in Preserving the effectiveness of Existing Antimicrobials

Himanshu Sekhar Pradhan
KIIT School of Public Health, KIIT Deemed to be University, Bhubaneswar, Odisha, India

Jyoti Prakash
KIIT School of Public Health, KIIT Deemed to be University, Bhubaneswar, Odisha, India

Mousumi Samal
KIIT School of Public Health, KIIT Deemed to be University, Bhubaneswar, Odisha, India

Sudhir Kumar Satpathy
KIIT School of Public Health, KIIT Deemed to be University, Bhubaneswar, Odisha, India

CONTENTS

12.1 Introduction

Antimicrobials are used to prevent and treat infections in humans, animals, and plants [1]. Antibiotics are one of the most successful discoveries in medical science. Since the discovery of penicillin in 1928 [2], several other antibiotics have revolutionised the treatment, control and elimination of many infectious diseases. Nevertheless, the effectiveness of antibiotics is hampered at this time by the increasing spread of antibiotic resistance (ABR). Although so many antibiotics have been developed over the years, many of them are no more effective to the pathogens. Since the development of resistance to penicillin in 1950, many bacteria, parasite, fungi and viruses have developed resistance to either single or multiple drugs. According to Global Antimicrobial Resistance and Use Surveillance System (GLASS) data, the rate of resistance to ciprofloxacin used to treat urinary tract infections varies from 8.4 percent to 92.9 percent for Escherichia coli and from 4.1 percent to 79.4 percent for *Klebsiella pneumonia* in countries reporting to GLASS. Fluoroquinolone resistance is widespread in *E. coli*, which is used to treat urinary tract infections [1]. The major concerns for tuberculosis treatment and control are 'multi-drug resistant tuberculosis (MDR-TB)' and 'extensively drug-resistant tuberculosis (XDR TB).' According to WHO estimates, in 2018, approximately 500,000 new cases of rifampicin-resistant tuberculosis (RR-TB) were found around the world, with the majority of these cases being 'MDR-TB'[1]. Drug resistance cases have also been reported in case of HIV (HIVDR), making the antiretroviral (ARV) drugs at risk of becoming partially or fully inactive. Artesunate-sulfadoxine-pyrimethamine failures in some countries have been reported due to development of resistance by the malaria parasite, Plasmodium falciparum to sulfadoxine-pyrimethamine, calling for new Artemisinin-based combination treatments (ACT). Candida auris-caused invasive fungal infections have developed resistance to fluconazole, amphotericin B, and voriconazole, as well as emerging caspofungin resistance[3].

The above facts reflect that antimicrobial resistance has become a reality and it is a silent pandemic looming on the horizon [4]. Among the Antimicrobial Resistance (AMR), the antibiotic resistance is the most prominent one. AMR is one of the top ten global public health threats that humanity faces today [1]. It is estimated that by 2050, yearly 10 million deaths could be caused by AMR [5]. To achieve the Sustainable Development Goals, urgent multi-sectoral action is required to combat AMR (SDGs) [1]. Antibiotics and other antimicrobial medicines are becoming ineffective as a result of the emergence of drug resistance, making infections increasingly difficult or impossible to treat, resulting in longer hospitalisation, higher treatment costs, and an economic burden on family and nation. Given the gravity of the AMR situation, the World Health Assembly adopted a global antimicrobial resistance action plan in May 2015, directing member states to develop and implement country-specific AMR action plans [6]. As per the global Tripartite (FAO, OIE and WHO) survey, national action plans for AMR are in place in 100 countries and 51 countries are in process of development of AMR plan [7] using One health approach. In view of the growing concern of ineffectiveness many existing antimicrobials, there has been call for development and accessibility of new antimicrobials in general and antibiotics in particular. However, development of new antibiotics has several constraints. It is resource intensive process which demands scientific research, innovation, funds and time. Drug companies are hesitant to develop new antibiotics due to scientific, regulatory, and economic considerations. Only a few countries, however, have addressed this issue by introducing financial incentive models to encourage antibiotic innovation [8]. It can take more than a decade to develop a new antibiotic, which can cost more than a billion dollars (850 million Euro). This financial and time investment must be safeguarded by implementing sustainable use measures that will extend antibiotic efficacy [9]. According to historical data, only one out of every five infectious disease drugs that reach the initial phase of human testing will be approved by the Food and Drug Administration (FDA). Only six of the 32 antibiotics in clinical development that target the WHO list of priority pathogens were deemed novel. Furthermore, a major issue is a lack of access to high-quality antimicrobials [1]. The new antibiotics are few and far.

In view of the global increase in the development of AMR, and newer drugs being few and far, the other alternative before the policy makers, researchers as well as implementers is to emphasize the preservation of the effectives of the existing antimicrobials. Among the Antimicrobial resistance, the antibiotic resistance (ABR) being the prominent one, further discussion in this chapter is focussed on the preservation of the effectiveness of existing antibiotics. In order to discuss further in perseverance of the effectiveness of existing antibiotics, it is necessary to understand the resistance development process of the bacteria to the antibiotics. The bacteria acquire resistance either by natural process or augmented by the several human factors. In case of natural process, the bacteria become resistant by mutation or by receiving the resistance genes from other bacteria through horizontal gene transfer. Bacteria uses various defense strategies (resistance mechanism) to resist the effects of antibiotics. Antibiotic resistance occurs naturally, but antibiotic misuse in humans and animals hastens the process [10].Several human factors accelerating the process of resistance development in bacteria are as follows with appropriate suggestions against these drivers of AMR for preserving the effectiveness of antimicrobials specifically antibiotics.

12.2 Irrational antibiotic use in the human sector

Without a doubt, the overall volume of antibiotic use and irrational antibiotic use, which is influenced by a variety of interconnected factors, are major contributors to the development of ABR. The drivers of irrational antibiotic use can be divided into two categories: 1. Irrational Antibiotic Use Among the General Public; and 2. Irrational Antibiotic Use Among Healthcare Providers.

12.2.1 Irrational Antibiotics Use Among the General Public

Irrational antibiotics use among the general public includes.

12.2.1.1 Poor Public Knowledge and Awareness

Antibiotic knowledge, attitudes, and beliefs are strong predictors of irrational antibiotic use [11]. Various studies conducted in different countries reveal that there is low awareness among general community regarding antibiotic resistance and its implication. Different study reports highlight higher volume of antibiotics use in low education and income groups as well as high income groups [12][13]. Raising public awareness is a neglected area, which needs to be given adequate attention in the action plan towards ABR containment.

12.2.1.2 Access to Antibiotics Without Prescription

Antibiotics available over-the-counter (OTC) without a prescription are common in many developing countries, which is one of the key drivers of irrational antibiotic use [11]. It is recognized that the OTC sale of antibiotics is a pathway for the emergence of AMR [14]. This may be due to inadequate access to proper diagnosis and investigation facilities [11]. Although the existing law prohibits the selling of antibiotics without the prescription still the practice prevalent in many countries especially in developing countries. Creating awareness among the dispensers, healthcare providers and the public to prevent the sale and access of antibiotics without prescription. Regular monitoring of the pharmacy outlets regarding compliance of existing law of dispensing antibiotics is required. A coordinated move with the help of various other ministries apart from health is necessary for compliance of the policy, roles and guidelines for the same. In many developing countries prescription of antibiotics by informal health care providers is rampant. Awareness creation of the informal healthcare providers regarding ABR as well as enforcement of treatment guidelines for antimicrobial use needs to be followed.

12.2.1.3 Self-medication

Self-medication of antibiotics is also prevalent in many part of the world, leading to incomplete, in-appropriate and overuse of antibiotics which facilitate development of ABR. Implications of self-medication should be included in all public awareness campaign on ABR.

12.2.1.4 Leftover Antibiotics

Leftover antibiotics from previous prescriptions/purchased antibiotics promotes the practice of self-medication. Therefore, dispensing exact number of antibiotics by the dispensers should be promoted. At the same time the patients/users should be educated not to use leftover antibiotics.

12.2.2 Irrational Antibiotics Use among Healthcare Providers

Irrational antibiotics use among healthcare providers include.

12.2.2.1 Antibiotic Use and Resistance: Healthcare Providers' Knowledge, Attitudes, and Perceptions

Prescribers are in charge of making decisions about whether or not to use antibiotics, as well as selecting the type of antibiotic. Therefore, their knowledge, attitudes, and perceptions regarding antibiotic use and resistance are likely to influence prescribing behaviour. It has been proposed that the attitudes and knowledge of physicians influence the quality of antibiotic prescribing as measured by clinical practise indicators. It has been proposed that physicians' attitudes and knowledge influence the quality of antibiotic prescribing as measured by clinical practise indicators. Steps to be taken to improve the education of the healthcare professionals at all levels both formal and Continuing Medical Education (CME) on the subject of AMR/ABR.

12.2.2.2 Inadequate professional education for Healthcare Providers

Inadequate antibiotic prescribing training in medical school and the early stages of clinical practise, as well as insufficient continuing professional development throughout careers, can all contribute to inappropriate antibiotic prescribing [11]. The WHO emphasises the significance of undergraduate education in prudent prescribing behaviour.. (World Health Organization 2012). The extent of the problem and the trend ; the strategies for containment and the role of health professionals in containing AMR/ABR should be adequately included in the curriculum of the formal education.

12.2.2.3 Pharmaceutical Promotion

Promotion of pharmaceutical is believed to escalate irrational prescribing and influences healthcare professionals' prescribing and dispensing behaviour [15]. Countries should ensure that adequate and effective methods are put in place to monitor the advertising of medicinal products.

12.2.2.4 Lack of Rapid and Sufficient Diagnostic Tests

One of the reasons for incorrect antibiotic prescribing and use is a lack of diagnostic tests that can quickly identify the pathogen and its antibiotic susceptibility profile, which can guide antibiotic prescribing at the point of care and reduce the need for broad spectrum antibiotics [15]. Provisioning should be made to improve availability and accessibility to diagnostic facilities and test for infectious diseases and antibiotics susceptibility test.

12.2.2.5 Patient-doctor Interaction

Doctor-patient interaction and communication as well as patient demands influence doctors to prescribe antibiotics.

12.2.2.6 Knowledge, Attitude and Perception of the pharmacists /personnel manning the pharmacy stores towards Antibiotic Use and Resistance

Pharmacists in community and hospital settings, as well as those who work in pharmacy stores as part of the dispensing process, should counsel patients on the safe, effective, and rational use of medications. This has been a neglected topic in the effort of containing AMR. Hence their knowledge, needs to be enhanced for complying standard dispensing practices [16].

12.3 Irrational use of antibiotics in food animals and agriculture sectors

Intensive farming has increased many folds especially in developing countries as a result of increased demand for animal protein resulting in antibiotic residues in animal-derived products and eventually antibiotic resistance. The antibiotics are used in animal sectors for the therapeutic, prophylactic and metaphylactic purpose. Other than these, antibiotics are also extensively used for growth promotion of animals, especially in poultry which is also a leading reason of misuse of antibiotics [17]. Antibiotic-resistant bacteria associated with animals can easily be transmitted to humans via food chains, and they can also be transmitted into the environment via animal waste.. According to studies, the global average annual consumption of antimicrobials per kilogramme of animal produced was 45 mg for cattle, 148 mg for chickens, and 172 mg for pigs [18].

Antibiotics have been used to control certain bacterial diseases of high-value fruit, vegetable, and ornamental plants since the 1950s. Oxytetracycline and streptomycin are the most commonly used antibiotics on plants. According to research, a subset of streptomycin-resistance genes found in plant-associated bacteria are similar to those found in bacteria isolated from humans, animals, and soil that contain transfer-proficient elements.The link between antibiotic use on plants and antibiotic resistance issues in human medicine, on the other hand, is being debated [19]. Strategies to reduce the extent of antibiotic use in animals are: prevention of disease and infection; implementing stringent hygiene practices; feed quality, and healthy environmental conditions to reduce stress; optimizing nutrition to enhance natural immunity; breeding genetically disease-resistant livestock [20]; holding education campaign for farm owners on workers to avoid irrational use of antibiotics and good farm management.

12.4 Other influencers of ABR

Other influencers of ABR are as follows.

12.4.1 Poor infection prevention and control practices in health care facilities and animal farms

Poor infection prevention and control (IPC) leads to spread of infections and resistant bacteria to patients and healthcare workers, which demands use of more antimicrobials specifically antibiotics. Improper animal farm management including hygiene and sanitation also increase the chance of contracting and spreading the infections. A formal infection control programme should be implemented in health care facilities, with dedicated infection control practitioner (ICP) to coordinate the programme. Hand hygiene, cleaning, and disinfection are examples of general infection control practices used in health care facilities to reduce infections and contain AMR [21]. Biocontainment measures should be implemented in animal farms to prevent the spread of infectious agents among farm animals or to keep infectious agents from leaving the farm [22]

12.4.2 Unhygienic living condition and poor personal hygiene

This leads to incidence of more infectious diseases, which results into more antibiotics use for treatment. Safe water and sanitation reduce the incidence of infectious diseases and its speared in the community. Infrastructure investment to improve hygiene and living conditions is easily justified as a tool for limiting the spread of antimicrobial resistance in communities worldwide [23].

12.4.3 Improper disposal of pharmaceuticals waste

Pharmaceutical waste has been found at trace amounts in drinking and surface water due to improper disposal , which may lead to development of resistant strain of bacteria. To address the problem of safe pharmaceutical waste disposal, a practical, environmentally sustainable approach to pharmaceutical waste management, with policies, roles and guidelines, as well as public awareness campaigns, are required [24].

12.4.4 Other enabling strategies to enhance the sustainability of the effectiveness of the existing antibiotics are as follows

• **Global and National AMR Action plan**: In May 2015, the World Health Assembly endorsed the Global Action Plan for AMR, which outlined five objectives for combating antimicrobial resistance and urged member countries to frame and implement AMR National Action Plans that are in line with the Global Action Plan.

• **Surveillance of AMR/ABR**: Surveillance is a critical tool for informing policy as well as infection prevention and control strategies. It is also required for assessing AMR spread and informing and monitoring the impact of local, national, and global initiatives. The World Health Organization (WHO) established the GLASS in 2015, the first global collaborative effort to standardise antimicrobial resistance surveillance. GLASS seeks to integrate data on the state of participating countries' AMR monitoring systems with AMR data for chosen priority bacteria that cause illnesses in people during its early implementation stage (2015-2019) [25]. Collecting AMR surveillance data is critical for
(1) defining the magnitude of the resistance problem,
(2) developing interventions to enhance the proper use of antimicrobial drugs, and
(3) reducing resistance selection pressure. The AMR surveillance must be a part of national and local level action plan for AMR.

• **Stewardship:** Antimicrobial stewardship (AMS) is described as "an organisational or healthcare-system-wide strategy to encourage and monitor the prudent use of antimicrobials in order to retain their efficacy. The antimicrobial stewardship programme (AMS programme) is a health-care strategy that promotes

the proper use of antimicrobials through the implementation of evidence-based interventions.(World Health Organization 2019a). Antimicrobial stewardship program (AMSP) helps clinicians to improve:
1. The quality of patient care;
2. Patient safety;
3. Reduced treatment failures;
4. Increasing frequency of prescribing appropriate therapy and prophylaxis.
5. Reduces antimicrobial resistance [26].
Each hospital/ health care facility should have antimicrobial stewardship programme with AMS committee to execute, monitor and evaluate the stewardship initiatives.

• **Alternative to antibiotics**: Infectious diseases have long been treated with medicinal plants. Plant-derived compounds like quinine and artemisinin have already been used to treat potentially fatal infectious diseases. Many ancient remedies have been scientifically proven to work. More work, however, is needed to standardise and reproduce plant-derived products. Clinical trials evaluating the efficacy of plant-derived products, adult plant uses, gemmotherapy, and the use of embryonic stem cells should be prioritised [27].

• **Research and Development for new antimicrobial, tools and vaccines**: More investment in research and development specifically to develop diagnostic tools and vaccination that would help in sustaining the effectiveness of existing antibiotics. Research and development also need more funds to develop new antimicrobial. Pharmaceutical companies should be encouraged for research and development by providing incentives.

12.5 Conclusion

Antimicrobial resistance, particularly antibiotic resistance, is a growing public health threat that requires immediate action around the world. Containment of ABR requires concerted multi-sectoral (including NGOs) action using one health approach. Sustaining the effectiveness of existing antimicrobials/antibiotics require strong action oriented interventions such as – public awareness campaign; education and training at all levels; improving education for healthcare professionals (both formal and CME); clinicians' support for educating patients; infection prevention and control measures; antimicrobial stewardship programme; surveillance and mapping information on ABR; public health-driven models of innovation; focussed research; innovative, affordable and easy-to-use rapid diagnostic tests; use of medicinal plants and monoclonal antibody showing effectiveness in treating infectious diseases; stronger regulations governing pharmaceutical promotion; strong political will to support for the development and implementation of national action plans with adequate funding. Review of government policy and action plans on AMR/ABR from time to time would help to modify the strategies to combat AMR/ABR. Monitoring and benchmarking are necessary to assess the progress of these interventions.

Bibliography

[1] WHO. Antimicrobial resistance, 2019.

[2] Susan Aldrich. Alexander Fleming Discovery and Development of Penicillin - Landmark - AmericanChemical Society, 1999.

[3] World Health Organization et al. Who global strategy for containment of antimicrobial resistance. Technical report, World Health Organization, 2001.

[4] KATE DODSON. Antimicrobial resistance: the silent pandemic.

[5] World Health Organization. Antimicrobial Stewardship Programme in Health-Care Facilities in Low and Middle Income Countries, 2019.

[6] Mark Woolhouse, Melissa Ward, Bram Van Bunnik, and Jeremy Farrar. Antimicrobial resistance in humans, livestock and the wider environment. *Philosophical Transactions of the Royal Society B: Biological Sciences*, 370(1670):20140083, 2015.

[7] World Health Organization et al. Cop24 special report: health and climate change. 2018.

[8] Ilinca A. Dutescu and Sean A. Hillie. Encouraging the Development of New Antibiotics: Are Financial Incentives the Right Way Forward? A Systematic Review and Case Study, 2021.

[9] Authors Christine Årdal, David Findlay, Miloje Savic, Yehuda

Carmeli, Inge Gyssens, Ramanan Laxminarayan, Kevin Outterson, John H Rex, Christine Årdal, Francesco Ciabuschi, Judith Hackett, Stephan Harbarth, Elizabeth Hermsen, Charles Knirsch, Nicole Mahoney, Nathalia Murillo, Ursula Theuretzbacher, Enrico Baraldi, Esther Bettiol, Taimur Bhatti, Simone Callegari, Claudie Charbonneau, Abby Colson, Mark Guthrie, Silas Holland, Karianne Johansen, Jostein Johnsen, Cecilia Kållberg, Carl Anderson Kronlid, Olof Lindahl, Ka Lum, Lene Martinsen, John McDonald, Steve McKeever, Annelie Monnier, Christopher Okhravi, Matthew Renwick, Kellie Ryan, Live Storehagen, and Liz Temkin and. DRIVE-AB Steering Committee members. 2018.

[10] WHO. Antibiotic resistance, jul 2020.

[11] Anna Machowska and Cecilia Stålsby Lundborg. Drivers of Irrational Use of Antibiotics in Europe. *International Journal of Environmental Research and Public Health*, 16(1), jan 2019.

[12] Jim O'Neill. Tackling drug-resistant infections globally: final report and recommendations. 2016.

[13] K. Henricson, E. Melander, S. Mölstad, J. Ranstam, B.S. Hanson, G. Rametsteiner, P. Stenberg, and A. Melander. Intra-urban variation of antibiotic utilization in children: influence of socio-economic factors. *European journal of clinical pharmacology*, 54(8):653–657, 1998.

[14] Gerard Porter, Anita Kotwani, Lovleen Bhullar, and Jyoti Joshi. Over-the-counter sales of antibiotics for human use in India: The challenges and opportunities for regulation:. *https://doi.org/10.1177/09685332211020786*, 21(2):147–173, jun 2021.

[15] Geoffrey K. Spurling, Peter R. Mansfield, Brett D. Montgomery, Joel Lexchin, Jenny Doust, Noordin Othman, and Agnes I. Vitry. Information from Pharmaceutical Companies and the Quality, Quantity, and Cost of Physicians' Prescribing: A Systematic Review. *PLOS Medicine*, 7(10):e1000352, oct 2010.

[16] Lori M. Dickerson, Arch G. Mainous, and Peter J. Carek. The pharmacist's role in promoting optimal antimicrobial use. *Pharmacotherapy*, 20(6):711–723, 2000.

[17] Christy Manyi-Loh, Sampson Mamphweli, Edson Meyer, and Anthony Okoh. Antibiotic Use in Agriculture and Its Consequential Resistance in Environmental Sources: Potential Public Health Implications. *Molecules : A Journal of Synthetic Chemistry and Natural Product Chemistry*, 23(4), 2018.

[18] Inge Van Loo, Xander Huijsdens, Edine Tiemersma, Albert De Neeling, Nienke van de Sande-Bruinsma, Desiree Beaujean, Andreas Voss, and Jan Kluytmans. Emergence of methicillin-resistant staphylococcus aureus of animal origin in humans. *Emerging infectious diseases*, 13(12):1834, 2007.

[19] Patricia S McManus, Virginia O Stockwell, George W Sundin, and Alan L Jones. Antibiotic use in plant agriculture. *Annual review of phytopathology*, 40(1):443–465, 2002.

[20] Thomas P. Van Boeckel, Charles Brower, Marius Gilbert, Bryan T. Grenfell, Simon A. Levin, Timothy P. Robinson, Aude Teillant, and Ramanan Laxminarayan. Global trends in antimicrobial use in food animals. *Proceedings of the National Academy of Sciences of the United States of America*, 112(18):5649–5654, may 2015.

[21] The Canadian Committee on Antibiotic Resistance. Infection Prevention and Control Best Practices, 2008.

[22] Drs. David R. Wolfgang Ernest Hovingh. Biosecurity - A Practical Approach, 2016.

[23] Brooke M. Ramay, Mark A. Caudell, Celia Cordón-Rosales, L. Diego Archila, Guy H. Palmer, Claudia Jarquin, Purificación Moreno, John P. McCracken, Leah Rosenkrantz, Ofer Amram, Sylvia Omulo, and Douglas R. Call. Antibiotic use and hygiene interact to influence the distribution of antimicrobial-resistant bacteria in low-income communities in Guatemala. *Scientific Reports 2020 10:1*, 10(1):1–10, aug 2020.

[24] Ranga Velagaleti, Philip K. Burns, Michael Gill, and James Prothro. Impact of current good manufacturing practices and emission regulations and guidances on the discharge of pharmaceutical chemicals into the environment from manufacturing, use, and disposal. *Environmental Health Perspectives*, 110(3):213–220, 2002.

[25] WHO GLASS. Global Antimicrobial Resistance and Use Surveillance System (GLASS), 2015.

[26] ICMR. Antimicrobial Stewardship Program Guideline, 2018.

[27] Mihaela Ileana Ionescu. Are Herbal Products an Alternative to Antibiotics? — IntechOpen, 2017.

13

Technology Solutions to AMR: Focus Paediatric Population

Preethi John

Chitkara Global Health Institute, Chitkara University, Punjab, India

Shweta Jindal

Piramal Swasthya Management and Research Institute, Pune, India

Deena Mariyam

Kempegowda Institute of Medical Sciences, Bengaluru, India

Honey Tandon

Global Health Student, Kings College, London, UK; Practicing Dietician (PGDDPHN), India, ex HOD (Dietetics) Max Hospitals and Columbia Asia, Gurugram, India

CONTENTS

13.1 Introduction

Antimicrobial resistance (AMR) has been effectively combated in the past by the use of technology. This chapter will explore the effects of AMR on the paediatric population and how technology could be used as an effective solution. The chapter is organised to provide an overview of issues of AMR in children. This is provided through the lens of the paediatric population, nutrition perspective as well as gender perspective. The review discusses the trend in ways nutrition impacts gender-specific health in children. Different upcoming technologies including Artificial Intelligence are reviewed to address these challenges. The chapter concludes with a call to action.

13.2 Risk of AMR starts from birth

The risk of antimicrobial resistance is borne by everyone right from the moment of birth. With shift to increase in institutional delivery this has tremendously seen a spike. Infections during pregnancy can have adverse effects and may cause infections in the newborn baby. Infections due to resistant bacteria can be severe and may require readmission, intensive and prolonged care, and facilities for culture and susceptibility testing. Mortality increases if initial therapy is inappropriate. This in turn impacts the cost of care including the treatment modalities. Introducing the use of antibiotics at a young age in children and repeated abuse of antibiotics lead to developing resistance against the microbes and further making the treatment of disease a challenge [1].

13.3 AMR in the paediatric population

Infants and children have immature immune systems that render them more susceptible to diseases, including those caused by drug-resistant microbes present in their surroundings. Research further suggests that there are a number of children staying in an unhygienic environment not suitable to promote good nutrition in developing countries due to various reasons that include financial and environmental [2]. Thereby, exposing these underprivileged children to antibiotics at an early age of life further increases the chances of AMR many-folds. Children living in low-resource settings with limited access to health services face an even greater risk of acquiring and falling sick due to drug-resistant microbes. Lack of safe water, sanitation, and hygiene, and inadequate infection control, further enable the spread of AMR. An increase in AMR can increase child mortality especially if resistance to treatments like antiretrovirals, antimalarials, and TB and antifungals are there. Malaria, TB, pneumonia, dysentery are the most common causes of childhood diseases and children being resistant to available treatment is a cause for great alarm. This will immediately increase childhood mortality and morbidity. Newborns, infants and children under the age of 5 are particularly prone to infections. Many babies who face non-infectious causes of respiratory distress are still prescribed antibiotics till the time their sepsis screen is reported negative. Thus, in retrospect, there is a large incidence of antibiotic overuse in this age group. While children need the full-course antimicrobials at the right time and when needed, it is also important that health workers and caregivers guard against AMR by avoiding mis- or over-use of antimicrobials.

13.4 Gender and nutrition: relevant for understanding AMR patterns

Time and again researchers have proved that nutrition has a long-term effect on human health particularly in children [3]. As the positive relationship between nutrition and health has been proved for ages, so is the impact of malnutrition [4]. There are many studies to state that malnutrition increases the risk of infections, morbidity, and mortality in children [5]. There is a direct and complex relationship between nutrition and infection. Malnourished children indicate children who do not receive optimum nutrition to their age. These children fall ill quite frequently and are exposed to antibiotics at an early age of life [6]. This trend was surprisingly observed higher in developing countries than the developed ones. For example, a research study as early as 2013 concluded that the evidence was not enough to continue with regular antibiotics in severe acute malnutrition (SAM), in particular addressing the areas having less prevalence of HIV [7]. Recently, studies suggested how gender bias could influence nutrition thus, overall affecting the health of children and adults [8]. To add on, the female child was neglected over the male child. Another study found that if many siblings of the same sex were born to parents, the latest one had poor health outcomes and received less attention [9], less focus on diet and health, and considered as more of a burden. However, the preference for the male gender always had supremacy as compared to the female children. Poor health outcomes increased the risk of infections in these children, in particular the female child. With the immune system in children still not fully developed this puts them at the risk of resistant bacteria. Several studies suggest that exclusive breastfeeding goes long way in providing immunity in children. Whereas, data suggests that the breastfeeding rate is four times lower in girl babies than to boy babies [10]. This increases the chances of a high infection rate in female children than male children and a higher chance of antibiotics being introduced at an early age of life [11]. A study concluded after many trials, where nutritional supplements were provided to the limited response on the growth and development of malnourished children, there were concerns over the sideeffects of processed food supplementation [12]. This is another huge area of concern where research is required. On the other hand, new studies are emerging that find that dietary intervention can reduce antibiotic resistance in the intestines of obese children. It further proved that gut bacteria can be managed by dietary interventions and can be useful in managing antibiotic resistance. However, the data is limited and requires further in-depth studies [13]. Resistant bacteria makes the effective treatment of common infections more and more challenging and children are especially vulnerable. Trials further suggested that the presence of AMR microorganisms in our agricultural food production systems and food chain is a potential route of exposure to everyone including children. To add on, the prudent use of antibiotics in animal foods adds to the chances of antibiotic resistance in meat consumers [14]. Along with the food, the hygiene and sanitation conditions where the food is grown and cooked as well as the hygiene of the surrounding environment did make a striking contribution to the poor health outcomes of the children [15] and further increasing their dependency to be treated with antibiotics. All of this evidence brings out numerous challenges in the paediatric population due to AMR.

13.4.1 Healthcare challenges due to AMR among paediatric population

Institutional deliveries, use of neonatal ICU, on one hand save lives; on the other, they are also points where there is an increased use of antibiotics. This reaches the children at a very early age. However, the data on this is very poor at the individual child level. This is due to absence of electronic medical records, self medications which prevent its reporting into hospital records. This necessitates good medical reporting and documentation practices. Protocols for AMR stewardship were skewed to the adult populations. It is only in recent years there has been an acknowledgement of the impact on paediatric population as well as need for AMR stewardship for paediatric population in a different manner. This will require action to be taken at appropriate levels within the healthcare provider facility as well as inter-provider and with the government providing appropriate regulation on the same. The ability to address the challenge of AMR requires different healthcare providers that child will meet across different age group of the child to work in a coordinated manner. However the doctors are prescribing their antibiotics without being sensitised to AWaRe guidelines. The most common causes of childhood mortality like malaria, respiratory infections and their treatment options have already shown the impact of children being resistant to common modalities. It is important that continuous professional education among the medical and healthcare fraternity are carried out to update. Nutrition implications highlight the vulnerability of the children at both ends of the spectrum – those who are malnourished as well as those who are obese. Both categories at a higher risk of AMR due to the need for antibiotics to treat both ends. Hence prevention and health promotion are also vital in AMR. This evidence also brings out the need for multisectoral coordination beyond the boundaries of a healthcare facility. The risk to nutrition increases as the food itself can become a conduit due to the use of antibiotics both in animals as well as in soil.

13.5 Technology solutions to address these AMR issues in children

A review of technology options highlights the possibility of addressing these challenges faced by the paediatric population due to AMR.

The diverse technology in AMR can be classified under the following headings: Database like CARD (comprehensive antibiotic resistance database) ;AWaRe Classification (Act Watch Reserve) Database Data and storage solutions; EMR and Laboratory technology; genome sequencing Apps Artificial Intelligence Solution for Physician Assistance in Diagnosis and Dosage

13.5.1 Database

Recognising the global problem posed by AMR, WHO published a global action plan on antimicrobial resistance which had a focus to improve quality via improving hospital antibiotic usage. WHO programme brought out AWaRe (Act, Watch, Reserve) guidelines. This provides a comprehensive database of all antibiotics classified under these three categories. These guidelines have been updated for the paediatric population. It is important that all providers access this database which is regularly updated. This is a necessity in all hospital stewardship programmes. There are several other databases that go into details of the resistance like the CARD(comprehensive antibiotic resistance database) database. Technology allows large datasets to be captured and reported in the form of a database.

13.5.2 Data and Storage Solutions including range of Hospital Management Information System

AMR has also brought out voluminous data. This could be in multiple forms. It could be patient data along with medical history as well as antibiotic medication prescription history. Technology today provides a lot of solutions for data management, data retrieval, data storage and reporting. This includes from computerised medical records evolved to electronic medical records and now electronic health records.

Neo4j is one such technology solution provider which facilitates data storage and management. It also provides a very user-friendly reporting system that is user-centric via the graph database. This is being used by the physician community to offer a patient the right antibiotic for the right diagnosis at the right dosage. The knowledge graph supports physicians to treat their patients in the most effective way by providing up to date information of antibiotics in a user-friendly manner.

AMR for R is a free open source and independent R package to simplify the anlaysis and prediction of AMR and to work with microbial and antimicrobial data and properties by using evidence based methods.

13.5.3 Laboratory Technology Solutions

This is another area where different technology is being developed to aid the work to accelerate the pace to combat AMR. Besides developments in diagnostic technology for quicker identification but also complex and higher technology for genome sequencing are also available.

13.5.4 Apps

There is a slew of apps that has come out that addresses various AMR challenges. While these may not be specific for the paediatric population, this technology shows what is possible via app as a technology for addressing AMR concerns. From the playstore on one's smartphone a range of apps can be accessed. A simple search in Google playstore revealed over 25 different apps. They revealed apps which brought out the organisational policy like "Standford

Guide" which brough their institute guidelines of antimicrobial therapy, or AIIMS antibiotic policy via the app "AIIMS antibiotic policy,"; or the apps titled "WHO Antibiotic Resistance," "Antimicrobial resistance app," or "The Antibiotic Guide," or "Antibiotics Flashcards," or "Antimicrobial stewardship," or "CliMic-Antimicrobial tool," or "ABS- Antibiotic Stewardship,".

13.5.5 AI based solutions

Over and above all other technology the emergence of Artificial Intelligence and the possibilities it offers will tremendously in addressing AMR with the paediatric population.

In the pediatric field, one of the first articles on the use of AI was published in 1984. SHELP was a computer-assisted medical decision making system. It aimed at diagnosing inborn errors of metabolism. Thirty years later, the IBM Watson platform was used to provide valuable diagnosis and treatment support of rare pediatric diseases at the Boston Children's Hospital, USA. Since then there has been a surge in publications on the use of AI in various fields of paediatrics. The interest for reducing AMR in paediatric populations stems from the tremendous scope it offers as a support for diagnostic decisions, gene expression profiling of children like for example with neuroblastoma or lymphoblastic leukemia. In paediatric infectious diseases, there are many strategies for the application of AI, from the development of new antimicrobial drugs to the correct diagnosis and management of infectious diseases.

Artificial intelligence methods can speed up the process of exploring optimistic antibiotic combinations.

Sequencing-based AI applications have been employed to study AMR. Collecting clinical data to build clinical decision support systems could help physicians monitor trends in AMR to increase the rational use of antibiotics. The advantage AI offers is the ability to handle large datasets. It is also possible to employ AI to design new antibiotics and synergistic drug combination investigations.

Deep learning approach within AI could mark a paradigm shift for future antibiotic discovery to design new antibiotics and optimize existing molecules.

13.6 Call to action

Technology development and deployment as a critical strategy in addressing paediatrics.

13.6.1 AMR challenges

Power of technology is critical to leverage to address the complex set of challenges brought by AMR for the paediatric population.

13.6.2 AI application in AMR

This is really the future of unravelling the complexity of AMR work and must be taken up in a phased manner to convert machine learning to deep mind.

13.6.3 Gender-disaggregated data on resistance are crucial

This will increasingly be a necessity to monitor the situation and help to track the effectiveness of policies in addressing the AMR threat.

13.6.4 Paediatric AMR Stewardship is key

This can be given greater emphasis with use of technology to support physicians in decision making as well as in online training on these vital issues.

13.6.5 Ensure children and their caregivers have equitable access to antimicrobials

In recognition of children's right to health requires governments to develop and implement policies and regulations to facilitate this. Observe the opportunity of World Children's Day as an opportunity to sensitive children and relevant stakeholders on ABR. Guided by the Convention on the Rights of the Child, recognition of children's right to health requires governments to develop and implement policies and regulations that ensure children and their caregivers have equitable access to antimicrobials, and to inform them on when and how to use the medicine. In addition, policies, regulations and protocols must be in place to prevent over prescription of antimicrobials by health professionals and to avoid over-the-counter sales. For example, reducing the incidence of infections and preventing malnutrition by encouraging breastfeeding and improving water and sanitation standards is key for reducing antibiotic use. To add on, preventing the development and spread of antibiotic resistance through prudent use of antibiotics in food animals by awareness amongst animal producers. However, this is an emerging topic to be considered, limited studies are available and a huge research gap is there in particular to gender and requires further studies in this area.

13.7 Conclusion

In conclusion, introducing antibiotics at an early age of life and repeated abuse of antibiotics leads to developing resistance against the microbes that make the situation further challenging [16]. Henceforth, the prevention of antibiotic resistance particularly for the paediatric population is a task that should involve many approaches and different sectors. It requires international cooperation collaboratively working in this field. The aim should be to prevent and control infections through the awareness and support of the developing countries to reduce the use of antibiotics at early ages of life. Continued efforts for robust National and Global initiatives need to be taken for identifying problems, resolving them in time including creating relevant policies and adopting them. Opportunities like World Antimicrobial Awareness Week, must be used to not only increase awareness of AMR but also encourage best practices among the public, health workers and policy makers. They also aim to bring to the forefront, children's rights to health. World Children's Day, which falls around the same period, is an opportunity to celebrate and reflect on collective efforts to ensure that children's rights, including their rights to good health and health care – are respected, protected and fulfilled. Right to health means being entitled to quality health care, including access to essential drugs and medicines, such as antimicrobials. It also means having a right to access appropriate health information, including on the proper use of antimicrobials and the risks associated with their mis- or overuse. Children must not be left behind in the global AMR response. Their unique needs and vulnerabilities must be addressed and their rights to health upheld. It is the responsibility of adults to protect them against this rising threat.

Bibliography

[1] Claire Roubaud-Baudron, Victoria E Ruiz, Alexander M Swan Jr, Bruce A Vallance, Ceren Ozkul, Zhiheng Pei, Jackie Li, Thomas W Battaglia, Guillermo I Perez-Perez, and Martin J Blaser. Long-term effects of early-life antibiotic exposure on resistance to subsequent bacterial infection. *MBio*, 10(6):e02820–19, 2019.

[2] Zalak Rameshbhai Matariya, Kaushik Kishorbhai Lodhiya, and Rajkumar G Mahajan. Environmental correlates of undernutrition among children of 3–6 years of age, rajkot, gujarat, india. *Journal of family medicine and primary care*, 5(4):834, 2016.

[3] Andrea Di Cesare, Ester M Eckert, Silvia D'Urso, Roberto Bertoni, David C Gillan, Ruddy Wattiez, and Gianluca Corno. Co-occurrence of integrase 1, antibiotic and heavy metal resistance genes in municipal wastewater treatment plants. *Water Research*, 94:208–214, 2016.

[4] Olaf Müller and Michael Krawinkel. Malnutrition and health in developing countries. *Cmaj*, 173(3):279–286, 2005.

[5] Judd L Walson and James A Berkley. The impact of malnutrition on childhood infections. *Current opinion in infectious diseases*, 31(3):231, 2018.

[6] Anjelique Schulfer and Martin J Blaser. Risks of antibiotic exposures early in life on the developing microbiome. *PLoS pathogens*, 11(7):e1004903, 2015.

[7] Gabriel Alcoba, Marko Kerac, Serge Breysse, Cécile Salpeteur, Annick Galetto-Lacour, André Briend, and Alain Gervaix. Do children with uncomplicated severe acute malnutri-

tion need antibiotics? a systematic review and meta-analysis. *PloS one*, 8(1):e53184, 2013.

[8] Maria Marino, Roberta Masella, Pamela Bulzomi, Ilaria Campesi, Walter Malorni, and Flavia Franconi. Nutrition and human health from a sex–gender perspective. *Molecular Aspects of Medicine*, 32(1):1–70, 2011.

[9] Kanika Goyal, Pragati Purbiya, Sandeep N Lal, Jaspreet Kaur, Pooja Anthwal, and Jacob M Puliyel. Correlation of infant gender with postpartum maternal and paternal depression and exclusive breastfeeding rates. *Breastfeeding Medicine*, 12(5):279–282, 2017.

[10] Edgar Manuel Vásquez Garibay, Elizabeth Guzmán Mercado, Alfredo Larrosa Haro, and Nelly Carolina Muñoz Esparza. Is there gender discrimination in full breastfeeding in mexico? *Nutrición hospitalaria: Organo oficial de la Sociedad española de nutrición parenteral y enteral*, 36(3):545–551, 2019.

[11] Frank I Scott, Daniel B Horton, Ronac Mamtani, Kevin Haynes, David S Goldberg, Dale Y Lee, and James D Lewis. Administration of antibiotics to children before age 2 years increases risk for childhood obesity. *Gastroenterology*, 151(1):120–129, 2016.

[12] André Briend, Peter Akomo, Paluku Bahwere, Saskia De Pee, Filippo Dibari, Michael H Golden, Mark Manary, and Kelsey Ryan. Developing food supplements for moderately malnourished children: lessons learned from ready-to-use therapeutic foods. *Food and nutrition bulletin*, 36(1_suppl1):S53–S58, 2015.

[13] Marina Sanchez, Shirin Panahi, and Angelo Tremblay. Childhood obesity: a role for gut microbiota? *International journal of environmental research and public health*, 12(1):162–175, 2015.

[14] Nikola Vidovic and Sinisa Vidovic. Antimicrobial resistance and food animals: Influence of livestock environment on the

emergence and dissemination of antimicrobial resistance. *Antibiotics*, 9(2):52, 2020.

[15] Annette Prüss-Ustün, Jamie Bartram, Thomas Clasen, John M Colford Jr, Oliver Cumming, Valerie Curtis, Sophie Bonjour, Alan D Dangour, Jennifer De France, Lorna Fewtrell, et al. Burden of disease from inadequate water, sanitation and hygiene in low-and middle-income settings: a retrospective analysis of data from 145 countries. *Tropical Medicine & International Health*, 19(8):894–905, 2014.

[16] Bilal Aslam, Wei Wang, Muhammad Imran Arshad, Mohsin Khurshid, Saima Muzammil, Muhammad Hidayat Rasool, Muhammad Atif Nisar, Ruman Farooq Alvi, Muhammad Aamir Aslam, Muhammad Usman Qamar, et al. Antibiotic resistance: a rundown of a global crisis. *Infection and drug resistance*, 11:1645, 2018.

14

Human resources: Capacity building for AMR Stewardship programme

Preethi John

Chitkara School of Health Sciences, Chitkara University, Punjab, India

Navneet Kaur Bains

Chitkara School of Health Sciences, Chitkara University, Punjab, India

Sandhya Vashisht

Project manager,Tech Mahindra SMART for Healthcare, Mohali, India

Keerti Bhusan Pradhan

Chitkara School of Health Sciences, Chitkara University, Punjab, India

CONTENTS

14.1 Introduction

Antimicrobial Stewardship is a rigorous program for containment of AMR in any healthcare setup and has been flagged as priority area for action by WHO [1]. It requires a multidisciplinary team to work on this [2]. Today this is incorporated in curricula of many undergraduate and postgraduate curricula [3]. However in majority of the places it is the inhouse training or continuous medical education which provides overview of AMR stewardship [3]. In India several measures were taken to roll out a strong AMR stewardship programme [4]. Indian Council of Medical Research rolled out a programme christened ASPIC Antibiotic Stewardship Prevention of Infection and Control [5]. Very detailed work on this has been done across the state of Kerala [6]. During the COVID pandemic the AMR stewardship programme was affected [7, 8]. There was early recognition that it needs to be addressed [9]. In India too there were several problems during COVID 19 on the stewardship programme [10]. This chapter outlines the human resources that constitute training that takes place in an antibiotic stewardship programme.

Methods review of literature was carried out with the different cadre of human resources that make up the AMR stewardship team in hospitals. Interview methodology with AMR stewardship team leads in three tertiary hospitals of Punjab, India brought out the functioning of the AMR team as well as the challenges faced of continuing this work during the pandemic period.

14.2 Hospital based Antimicrobial Stewardship Team

Antibiotic Stewardship Committee is a multidisciplinary team in any tertiary care. Tertiary care hospital comprises two teams: the core team as well as the monitoring team.

14.2.1 Core Team

This team also referred to as the ground team comprises the primary caregivers on the ground. This multidisciplinary team comprises

- Head Infectious Diseases Consultant/Head Critical Care/ Head ICU (Heads the core team and takes decisions related to treatment)
- Head Internal Medicine Consultant (Takes decision related to treatment)
- Head Microbiology Consultant (Closely investigates diagnostics and labs)
- Nursing Administration (Guides and are responsible for training Nurses)
- Head Infection Control Nurse (Gives inputs related to Infection Control)

14.2.2 Monitoring Team

This team works closely with the Core team but basically monitors the Stewardship Program (Table 14.1.). This multidisciplinary team comprises of the

1. Head Pharmacology (This person is very often considered the single point of contact (SPOC) for AMS related data of the Hospital and main responsible person of AMS team, arranges lectures and seminars for the area ground staff)
2. Medical Director (MD)/ Medical Superintendent(MS) (Monitors the team and main decision taking power lies with the MD/MS)
3. Clinical Pharmacologists (They directly report to Head Pharmacology, active on ground members of monitoring team, audits and work closely with the ground staff, collects and presents antibiotics data in all meetings)

This study revealed that it is included in most curricula of degree programme. However, it is alarming that it is not included in the curricula of dentists. There is enough evidence that antibiotic prescription via the dentist route is among the highest. It is also not reflected in the curricula provided by Pharmacy Council of

TABLE 14.1
AMR Stewardship included in curricula of undergraduate and postgraduate education in India

Degree	Antimicrobial /antibiotic stewardship program whether included in curriculum	Source
MBBS	Yes	https://www.nmc.org.in/wp-content/uploads/2020/01/UG-Curriculum-Vol-I.pdf
BDS	No	https://dciindia.gov.in/Rule_Regulation/Revised_BDS_Course_Regulation_2007.pdf
BAMS	No	https://www.ccimindia.org/ayurveda-syllabus.php
BHMS	No	https://www.tnmgrmu.ac.in/images/Syllabus-and-curriculam/syllabus-and-regulations-indian-medicine-homoeopathy/BHMS-REGULATION-19032015.pdf
BSc.Nursing	Yes	https://www.pdm.ac.in/wp-pdmu/uploads/2016/10/B.Sc_Nursing_Syllabus_2019-20.pdf
Allied health Professions	Yes(except optometry)	https://main.mohfw.gov.in/Organisation/departments-health-and-family-welfare/pms/model-curricula-allied-and-healthcare-streams-0
MBA Healthcare management / MPH	Yes	
Pharmacy	No	

India. This again is surprising as there is enough evidence of over the counter medications including antibiotics being sold.

14.3 Hospital based Training of Staff in AMR Stewardship

Hospitals and Stewardship Committee have made Standard Operating Procedures (SOP's) and processes to review the ongoing prescription and flow of antibiotics in the Hospital. Any new team member who joins in any of the teams are trained for Basic working of the Committee on ground for 1-2 months via various trainings.

14.3.1 Human Resource Training

The human resource trainer trains every new joiner on the basic working of the Hospital which includes lectures from major teams and departments of the Hospital for, e.g., Quality Department, Infection Control, Lab and Diagnostics, Security, etc.

14.3.2 On the job training

The person is trained on ground by the immediate senior on the basic SOPs of that particular area or department for around 1 to 2 months.

14.3.3 The training for AMS Committee covers

The training for AMS Committee covers:

14.3.3.1 Proper filling of the Antibiotic Justification/Administration Form

Whenever any treating Physician or Consultant prescribes any antibiotic to the patient, he/she needs to fill the antibiotic Justification/Administration Form where he justifies the chosen Antibiotic its dosage and frequency along with the route of administration, keeping in mind the current infection and investigations. Following after few days of administration depending upon the current investigations and Infection the Physician escalate/ deescalate or change the antibiotic along with justification each step in the form.

14.3.3.2 Familiarity with the Restricted Antibiotic List

Few antibiotics come under high end or restricted list where their use is restricted and limited. This list is made by the Head Pharmacology in conjunction with the Core team and is circulated to all the concerned areas from the desk of Medical Director. From time to time this list is updated depending upon the suggestions from Core team, govt. guidelines or the Corporate /Head office of the Hospital. But the AMS Team do not follow these guidelines blindly as they carefully take the decisions depending upon the condition and type of patients in their hospital. If the Consultant feels the need to prescribe such antibiotics, he also needs to take permission from AMR Committee along with the approval from the Medical Director/ Medical Superintendent of the Hospital by the Justification documents along attached. This approval is written and documented. The Committee can also suggest the other potential alternatives for the same.

14.3.3.3 Audit Process

This training is done by the Clinical Pharmacologists of the monitoring team who are specifically hired for AMS Team with the educational qualification of Doctor of Pharmacy in their educational background which is a degree programme of 6 years. The audit process is done daily and the daily auditors check the compliance of the Hospital staff in various areas to the SOP's formulated by the AMR Stewardship team. They daily check the Justification/ Administration forms of all the patients under antibiotics along with the prevailing conditions and the investigations of the patient. They review the current dosages indication routes and duration of the Antibiotics and intervenes when necessary. They regularly review the old and new investigations of the patients to suggest Consultants to increase decrease or deescalate the antibiotics. This team also checks for the prescribed antibiotics for the other alternatives as first or second line of drugs. The complicated cases are discussed by the team with the treating Consultants.

14.3.4 Continuous Education

Continuous Education includes:

14.3.4.1 Weekly Medical Meetings

All the stakeholders attend these meetings weekly in the Boardroom of the Hospital for proper learning and discussions. These stakeholders include The Medical Director/Medical Superintendent of the Hospital, Medical Coordinators (who work under MD/MS in Med. Admin team), Doctor in charges of all areas including Wards, Critical Care units and other areas like Step down areas of ICU's, Nursing In charges of all areas, Pharmacology Head and AMS team along with Medical Records Head. In these meeting the data related to antibiotics and narcotics is discussed along with the other quality audit data. In antibiotics specifically the AMS Team present the data showing the missed doses, wrong doses, incomplete prescriptions and documentation related to each area. The reasons for the same are discussed and remedial measures are taken against the errors. Incomplete justification forms or documentation are also discussed from case to case and for each area of hospital (wards, ICU and Step downs)

14.3.4.2 Twice monthly AMS team meetings

Twice monthly AMS team meetings includes:
Seminars/Lectures/ Projects: - The AMS team members arrange and organize Seminars/Lectures and workshops for the continuous learning of all the members. The AMS team usually defines in the policy the number of lectures or awareness programs to be done in 6 months or 1 year. If need arises the policy can be changed. The Head Pharmacology along with Clinical Pharmacologists arrange these lectures for the ground staff for continuous learning.

WhatsApp Group: - For quick and close communication regarding the Antibiotics and the Stewardship program members communicate through the WhatsApp Group where all the team members along with the Hospital Management the part of the conversations.

14.3.5 Inter-Hospital Learning

Hospital Stewardship Committee members are regularly in touch with the other Stewardship Committees of the other hospitals. The SPOC for these arrangements is Head Pharmacology / Head Microbiology. In Fortis Mohali the SPOC is Head Microbiology. Government has appointed a Nodal Officer in each government hospital for Antimicrobial Stewardship and guides them to work closely with the AMS teams of tertiary care hospitals. Teaching Government hospitals like PGIMER and GMCH 32 are actively collaborating with private tertiary care hospitals to work for the program by closely monitoring the small nursing homes and hospitals of that area. They regularly engage and communicate through workshops and seminars on the same. They exchange the good practices of the hospitals.

14.3.6 ICMR Projects/ Government Projects

The team members actively participate in the government organized Projects and share the related data with the government. These projects provide opportunities for all to learn at the State and Interstate levels. Government is planning to link Diagnostic Stewardship to AMR Stewardship so that both committees can work hand in hand.

14.4 Stewardship Program during COVID 19

COVID 19 impacted in multiple ways the smooth functioning of the stewardship programme which was set up in the hospital and running smoothly.

14.4.1 Normal functioning of the hospital was disrupted

During the pandemic times the normal functioning of the hospital was impacted. Emergency admission of patients increased to such a large extent that OPDs had to be shut down to look after these patients. As admissions due to COVID 19 related to issues surged, other hospital admissions almost came to a halt. The focus was to save lives of people in very challenging conditions and constrained resources as well as in an atmosphere of tremendous fear. This in turn affected the functioning of all regular committees including the stewardship programme in hospital.

14.4.2 Standard Operating Protocols (SOPs) for COVID 19 patients

The SOPs of the hospital for patient care had to be immediately updated. The Antimicrobial Stewardship Programme of the Hospital was also affected.

14.4.3 Roles and responsibilities of staff were redefined

Roles and responsibilities of the staff had to be changed in a short span of time to cater to the immediate challenges thrown up by COVID 19. During this period, staff also came down with COVID 19 which further affected the availability of hospital personnel. This affected the regular functioning of the hospital. Audits carried out by the stewardship team were affected due to their not being considered as a high priority as well as team members unavailability due to illness or deputation on some other role. The need of Critical Care Consultants abruptly increased. Increasing bed capacity in the hospital was impacted by the the inadequacy of trained human resources in healthcare. There was a need for experienced staff which was limited. All those who were in training, internship were permitted to be not available in hospital from safety perspective. This group also was neither experienced or not trained to handle the needs of the COVID era.

14.4.4 Medical documentation was hampered

Patient medical records are considered an important part of the patients, medical history. This is a vital aspect for a good stewardship programme. As the pandemic progressed self medication and over the counter prescriptions skyrocketed. Several smaller hospitals closed, and due to the scare of visiting hospital, home delivery options of medicines became the preferred route to self treat. During the national lockdown in India not even chemists were allowed to be open during a period of time. Out of fear and due to misinformation which spread via social media patients were involved in self-diagnosis and self-medication. There were messages urging people to improve their immunity and AYUSH was also given a lift including by the government. All of this impacted the normal functioning of the stewardship program as antibiotics brought directly from the chemists without a prescriptions was not captured through the stewardship programme. Home healthcare surged with only the need for serious patients to be seen in hospitals. The rest could be seen via telemedicine and managed at home. A large number of patients were quarantined and isolated in places other than hospitals like in homes, hotels or other makeshift facilities which came up. Stewardship programme was not part of any of these temporary facilities. Even though each patient is different as medical history was not available, a chain of diagnostics did not take place, a standard prescription including of antibiotics was made for all. Due to lack of EMR medical documentation more often than not was seriously hampered and data vital for the stewardship programme was not captured. Different states had different rules in India; however, several governments provided the COVID positive home isolated patient with a COVID kit. This also included antibiotics. This reached the patients in their homes as well as in makeshift facilities. Again this was not accounted for, but the lack of knowledge about antibiotics uses in the majority of the population again worsened the Stewardship program. Those who were in no need of antibiotics were provided antibiotics though without any prescription about how to use them. But the major problem was giving a single type of antibiotic to all irrespective of age and need was a major failure to the Stewardship Program.

14.4.5 Stewardship team–functioning during COVID

Stewardship team continued to function during COVID. In several hospitals, two separate functioning AMS audit teams were made. One team looked after the COVID patients. The other team looked after all other categories of patients, which had considerably decreased during this period. However, the staff in both categories were severely constrained on time and this very often fell down in priority.

Priority and time availability did weekly meetings happen- Weekly Medical Meetings and Twice monthly AMS team meetings were happening Online as other meetings. But the Core members were majorly involved in saving sick COVID patients. The output of these meetings was decreased.

Training - Physical training was not possible due to physical distancing requirements, lockdown and other COVID related measures. The training was converted to an online format. However the members in looking after patient care that they very often missed online training or could not attend it.

Daily audit by the pharmacists: The team that was put on the audit for example, the pharmacy staff were not part of the pharmacy of the hospital. They were hired only for the purpose of working on the stewardship work. The separate role definition made it possible for this work to continue. However. Due to COVID, there were several changes which happened. Rounds were found to be difficult alongside incorporation of COVID related compliance in terms of adherence to personal protection equipment. Also large areas of the hospital was cordoned off to provide increased facility and space for isolation wards and for those who were positive and required to be quarantined. The daily audits were in place by both teams for COVID and Non COVID patients. However limited staff and time during rounds made the output questionable.

14.5 Call to action

“nobreak

14.5.1 Paediatric antibiotic stewardship program to be distinguished from adult stewardship program

Recognising the fact that children are high users of antibiotic it is very important to recognize that adult and paediatric are two aspects of the stewardship programme. There must be a balance between adults and paediatrics. Besides the paediatric department the children can be admitted via NICU, ICU, Emergency admission, orthopeadics, dentistry, paediatric cardiology and dermatology [11].

14.5.2 EMR to be strengthened in view of makeshift hospitals

EMR solutions to be available and medical documentation not to be compromised . Data capture of prescriptions is very vital to the stewardship programme [12]. This needs to be scaled up and will require technological solutions to be in place.

14.5.3 AMR Stewardship programmes to be strengthened to withstand shocks

As buildings are built to withstand earthquakes the hospital stewardship committees must reflect on how do they make their committees crisis and pandemic proof [13]. The robustness of the work needs to be built from several angles. First is within the hospital and the second is inter hospital stewardship functioning and the third is the role of the governmental regulations to promote stewardship function.

14.5.4 Culture change among healthcare leadership to incorporate technology and ensure data flow for stewardship

Technology and healthcare are twin sides of the future of improving health outcomes including addressing the challenges of antimicrobial resistance as well as AMR stewardship. However human resources in healthcare have been very reluctant to uptake technology. This attitude towards resistance to incorporation of technology needs to change in the hospital workflow. Healthcare leadership must influence the organizational culture to allow for this.

14.5.5 Interlinkage of antibiotic stewardship programmes to with diagnostic stewardship

It is very important the intersection between diagnostic stewardship and antibiotic stewardship is recognized. This must be interlinked and the policy of India needs to highlight and incorporate this as they make regulations to make the stewardship programme stronger.

14.5.6 Curricula to incorporate stewardship

All curricula for healthcare and medical personnel must include not only about antimicrobial resistance but also information about the role of antibiotic stewardship. This is not only critical for India but equally across the different countries of the developing world. It is in this context that PREVENT IT project of Erasmus Plus rolled out across specific universities in India and beyond is of such utmost significance.

14.6 Conclusion

This chapter has brought forward the urgent need to strengthen ABR stewardship in education of health and medical personnel both at the undergraduate and postgraduate level. At the same time, the inservice training in hospitals can leverage technology to ensure continuity for the programme even in the wake of a pandemic. This is the right time for concerted action by all across the board to reflect on strengthening the ABR stewardship programme by building human resource capacity of healthcare by employing multiple strategies.

Bibliography

[1] WHO Diagnostic Stewardship. A guide to implementation in antimicrobial resistance surveillance sites. *World Health Organization*, 2016.

[2] Susan Rogers Van Katwyk, Sara L Jones, and Steven J Hoffman. Mapping educational opportunities for healthcare workers on antimicrobial resistance and stewardship around the world. *Human Resources for Health*, 16(1):1–18, 2018.

[3] Md Anwarul Azim Majumder, Sayeeda Rahman, Damian Cohall, Ambadasu Bharatha, Keerti Singh, Mainul Haque, and Marquita Gittens-St Hilaire. Antimicrobial stewardship: Fighting antimicrobial resistance and protecting global public health. *Infection and drug resistance*, 13:4713, 2020.

[4] Jaya Ranjalkar and Sujith J Chandy. India's national action plan for antimicrobial resistance–an overview of the context, status, and way ahead. *Journal of family medicine and primary care*, 8(6):1828, 2019.

[5] Sujith J Chandy, Joy Sarojini Michael, Balaji Veeraraghavan, OC Abraham, Sagar S Bachhav, and Nilima A Kshirsagar. Icmr programme on antibiotic stewardship, prevention of infection & control (aspic). *The Indian Journal of Medical Research*, 139(2):226, 2014.

[6] Sanjeev Singh, Esmita Charani, Sarada Devi, Anuj Sharma, Fabia Edathadathil, Anil Kumar, Anup Warrier, PS Shareek, AV Jaykrishnan, and K Ellangovan. A road-map for addressing antimicrobial resistance in low-and middle-income countries: lessons learnt from the public private participation and co-designed antimicrobial stewardship programme in the state

of kerala, india. *Antimicrobial Resistance & Infection Control*, 10(1):1–9, 2021.

[7] Shamshul Ansari, John P Hays, Andrew Kemp, Raymond Okechukwu, Jayaseelan Murugaiyan, Mutshiene Deogratias Ekwanzala, Maria Josefina Ruiz Alvarez, Maneesh Paul-Satyaseela, Chidozie Declan Iwu, Clara Balleste-Delpierre, et al. The potential impact of the covid-19 pandemic on global antimicrobial and biocide resistance: an amr insights global perspective. *JAC-antimicrobial resistance*, 3(2):dlab038, 2021.

[8] Bindu S Mayi, Manda Mainville, Rida Altaf, Michelle Lanspa, Sheel Vaniawala, Thomas A Ollerhead, and Aarti Raja. A crucial role for antimicrobial stewardship in the midst of covid-19. *Journal of microbiology & biology education*, 22(1):ev22i1–2285, 2021.

[9] Jonathan Youngs, Duncan Wyncoll, Philip Hopkins, Amber Arnold, Jonathan Ball, and Tihana Bicanic. Improving antibiotic stewardship in covid-19: bacterial co-infection is less common than with influenza. *Journal of Infection*, 81(3):e55–e57, 2020.

[10] PS Seethalakshmi, Oliver J Charity, Theodoros Giakoumis, George Seghal Kiran, Shiranee Sriskandan, Nikolaos Voulvoulis, and Joseph Selvin. Delineating the impact of covid-19 on antimicrobial resistance: An indian perspective. *Science of The Total Environment*, page 151702, 2021.

[11] Patricia McGettigan, Peter Roderick, Abhay Kadam, and Allyson M Pollock. Access, watch, and reserve antibiotics in india: challenges for who stewardship. *The Lancet Global Health*, 5(11):e1075–e1076, 2017.

[12] Gaud Catho, Marlieke De Kraker, Brigitte Waldispühl Suter, Roberta Valotti, Stephan Harbarth, Laurent Kaiser, Luigia Elzi, Rodolphe Meyer, Enos Bernasconi, and Benedikt D Huttner. Study protocol for a multicentre, cluster randomised, superiority trial evaluating the impact of computerised decision

support, audit and feedback on antibiotic use: the computerized antibiotic stewardship study (compass). *BMJ open*, 8(6):e022666, 2018.

[13] Malcolm B Gillies, David P Burgner, Lorraine Ivancic, Natasha Nassar, Jessica E Miller, Sheena G Sullivan, Isobel MF Todd, Sallie-Anne Pearson, Andrea L Schaffer, and Helga Zoega. Changes in antibiotic prescribing following covid-19 restrictions: Lessons for post-pandemic antibiotic stewardship. *British journal of clinical pharmacology*, 88(3):1143–1151, 2022.

15

Combating AMR- One Health Approach

Dona Boban

Amrita Institute of Medical Science (AIMS), Kochi, India

Kiranjeet Kaur

Chitkara School of Health Sciences, Chitkara University, Punjab, India

Sanjeev K. Singh

Amrita Institute of Medical Science (AIMS), Kochi, India

CONTENTS

15.1 Introduction

Antibiotics were rightly termed as '"miracle drugs" in the last century. The discovery of antibiotics brought about considerable decrease in the deaths and disabilities due to infections. But the downside was the misuse of antibiotics not only in the human sector, but also in the animal and agriculture sector. Globally, more than 70% of antibiotics are used in animals for growth promotion and to prevent infections. The misuse of antibiotics exerts a selection pressure (Figure 15.1) on bacteria and acts as the perfect niche for the bacteria to become resistant. Ultimately, the antibiotic fails to cure the infections that they once did leading to antimicrobial resistance (AMR).

Antimicrobial resistance (AMR) is a global issue warranting the immediate attention from all sectors. The consequence is not only limited to mortality and morbidity. In fact, AMR has a significant impact on the global economy and security. Currently, AMR contributes to 7,00,000 deaths every year globally and it is predicted that by 2050, AMR will cause 10 million deaths every year and a 100 trillion US dollars loss in the world economy [1]. A recent review article has shown that around 4 million deaths are associated with bacterial AMR globally only in the year 2019. The drivers of AMR are misuse of antimicrobials in humans, animals and agriculture, poor infection control, run off waste from intensive agriculture, environmental contamination, poverty, poor sanitation and movement of infected humans and animals. AMR leads to long hospital stays, expensive treatments, increased side effects and complications and increased toxicity in humans. The consequences are similar in animals also.

AMR is not limited to a single region or country. The resistant bacteria and genes easily spread across the countries through humans, animals, food products, soil, water to name a few. The spread of AMR is better understood through a concept termed "One health." One health can be defined as "the integrative effort of multiple disciplines working locally, nationally, and globally to attain optimal health for people, animals, and the environ-

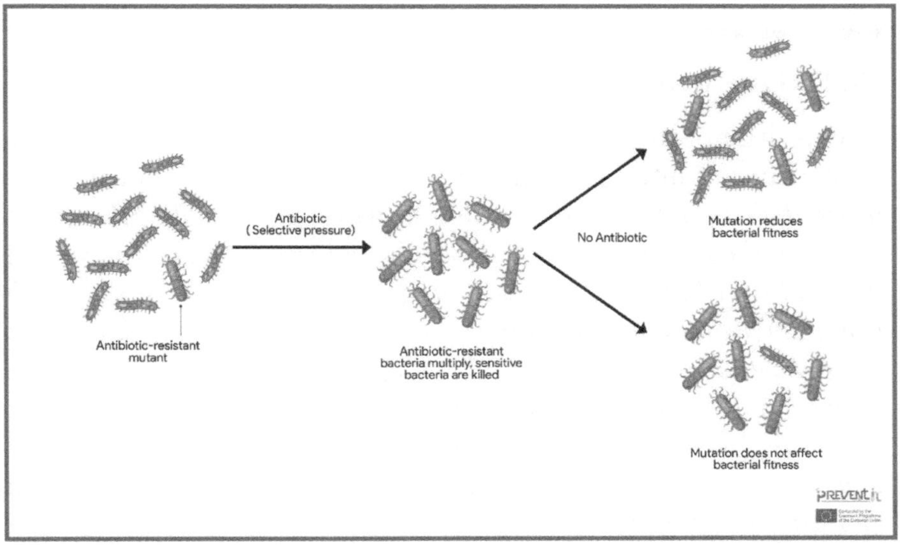

FIGURE 15.1
Emergence of antibiotic resistance.

ment" [2]. One health concept is not new. In fact, it was known from 400 BC onwards. The book "On Airs, Waters and Places" by Hippocrates refers to this concept. Throughout the history of medicine, many individuals contributed to one health like Rudolf Virchow, Louis Pasteur, Robert Koch, and Calvin Schwabe. One health recognises that human, animal, and environmental health are interlinked. The spread of AMR can be well explained through one health concept because antibiotics are given or consumed by humans, animals, and plants. The use of these antibiotics in any one domain has a definite impact on other domains as well. Additionally, the rapid population increase in developing countries makes this interconnection more impactful. Further, as humans expand to newer habitats, they come in close contact with wild animals, thereby increasing the risk of exposure to infections including the resistant ones in humans as well as animals. Therefore, action on prudent use of antibiotics in animals or humans alone will not lead us anywhere. Similarly, one health concept is imperative to prevent and control AMR because addressing antibiotic use across all sectors is the primary step in this direction.

15.2 One health

One health approach recognises that humans, animals, and the environment are intricately linked to each other. It is defined as "*an integrated, unifying approach that aims to sustainably balance and optimize the health of people, animals and ecosystems. It recognizes the health of humans, domestic and wild animals, plants, and the wider environment (including ecosystems) are closely linked and inter-dependent. The approach mobilizes multiple sectors, disciplines, and communities at varying levels of society to work together to foster well-being and tackle threats to health and ecosystems, while addressing the collective need for clean water, energy and air, safe and nutritious food, taking action on climate change, and contributing to sustainable development*" [3]. Basically, it is a new term with ancient roots.

This concept helps in understanding the concerns and challenges spanning across various domains, the full spectrum of diseases control and promotes health of all sectors. The world is becoming more complex as all domains are intricately linked to each other. Any action or intervention in any one domain has an impact on all other sectors. This can be easily identified in various situations like food security and safety, zoonoses, water safety, plant health, climate change, etc. For example, climate change has a direct effect on infection rate. It is known that clean air, clean water and extreme weather events are important for plants and agricultural products. So much so that it is expected to cause 250,000 additional deaths per year between 2030 and 2050, with 144,000 due to viral diarrhoea and malaria and 95,000 due to childhood undernutrition [4]. Among these situations, the prime example is AMR.

15.3 One health and AMR

AMR is considered a "One Health subject." Both the cause and solution of AMR are directly linked with the interrelationship be-

Antibiotics- The drugs used to treat bacterial or fungal infections

Antibiotic residues- The left over (metabolites) found in trace amounts in any edible portion of the animal product.

Antibiotic genes- The genes responsible for providing resistance to the microbes

Superbugs- The microbes which are resistant towards more than one antibiotic

FIGURE 15.2
AMR domains.

tween humans, animals, and the environment, and thus must be addressed at all the three domains globally. It is imperative to understand what the root cause of the AMR is to strategically plan and solve the problem. Thus, first one needs to understand more on how antibiotics are being used and thus entering the one health system.

15.3.1 Use of antibiotics

Antibiotics are widely used among humans, animals, and plants. In humans, those are used for prophylaxis and treatment of infections. Antibiotics are indispensable during routine surgeries, cancer chemotherapy, organ transplants, dialysis and treatment of inflammatory diseases. In animals, those are primarily utilised for promoting growth, mass prophylaxis and for therapeutic purposes. Most of the antibiotics produced globally are used in animals rather than humans. For example, data shows that in the USA, 80% of antibiotics produced are used in animals, and out of these 70% are critically important to humans [5]. In plants, antibiotics are used for therapeutic purposes, to reduce growth of bacterial pathogens, for seed treatment and in floriculture [6, 7].

Any use of antibiotics in humans and animals releases antibiotic residues into the environment through the excreta. It has been calculated that 75-90% of antibiotics consumed are released to the surrounding environment unmetabolized [8]. Additionally, any antibiotic use exerts a selective pressure in the bacteria present in the human and animal gut. Similarly, antibiotics in the environment exerts selection pressure on bacteria present in soil and water. In this situation human and animal gut, soil, water act as a factory for the generation of superbugs and continuously release such bacteria into the surrounding environment. To learn more about the emergence and spread of AMR and how it is linked to the one health concept, let's investigate each domain separately (Figure 15.2.).

15.3.2 Human domain

Humans release antibiotic resistant bacteria and genes generated in their gut and antibiotic residues in the form of excreta. If not treated appropriately, these resistant bacteria and antibiotic residues reach water channels and soil. Further, such soil and water is used in plant farming. Thus, superbugs and antibiotic residues get transferred from the agricultural sector to humans through consumption of vegetables and fruits grown in the farm or through drinking water. Humans can also acquire the resistant bacteria and genes from improperly cooked animal products as well. The spread of AMR in humans is depicted in the image (Figure 15.3) given below.

15.3.2.1 Risk factors

The vicious circle is further perpetuated through many anthropogenic factors. These include population rise, immunity status, globalization, and poverty.

1. Population rise: The global population is increasing exponentially over the years. Most of the population growth is observed in the developing nations, especially in the peri urban areas. This leads to the creation of slums with compromised hygiene and sanitation in these areas. Emergence of infectious diseases including the resistant ones is the expected outcome in such a scenario. Lack of public health facilities to identify and respond to these threats makes it riskier in developing countries.

2. Immunity status: In developed countries, the segment of

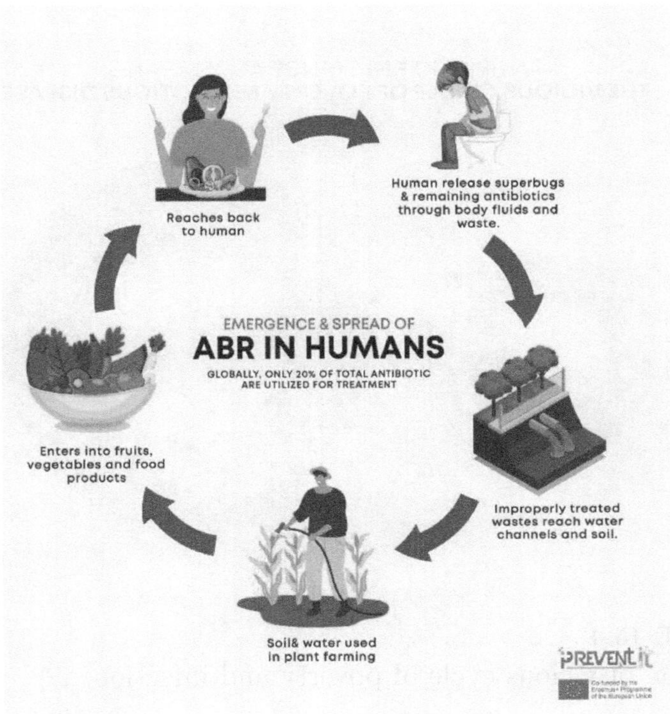

FIGURE 15.3
Spread of antibiotic resistant bacteria and antibiotic residues in humans.

elderly people is increasing, who are at higher risk of immunocompromised situations. Similarly, the cohort of people undergoing cancer chemotherapy, organ transplants, and immunocompromised diseases like AIDS are increasing globally. These cohorts are highly susceptible to infections including the resistant ones.

3. Globalization: Currently, the world has been reduced to a "global village." Global trade and commerce increased exponentially in this era. People are moving across the globe frequently. The relocation of people due to changing economy and job situations, movement of refugees due to political unrest are common in the present era. These movements merged the global space, leading to increased interactions among and between people, animals,

FIGURE 15.4
Depiction of vicious cycle of poverty and infection.

animal products and environment.

4. Poverty: Poverty also plays a major role in increasing the AMR infections. People who are trapped in poverty have reduced quality of life and poor health. This makes them more susceptible to acquiring infections including the resistant ones. If infected, they do not have necessary resources to treat the same, which leads them to more misery (Figure 15.4). Most of the poor people, especially in Southeast Asia, Africa and India depend on animal agriculture for their day-to-day living [9]. Due to their dependence on animals and animal products for their livelihood, they are highly susceptible to zoonotic infections.

15.3.3 Animal domain

More and more animals are becoming pet animals. Exotic animals are in demand and their trade has increased across the globe. The increasing need for animal protein has led to more intensive animal farming compromising the hygiene and safety. Animals act as vehicles (resistant bacteria can multiply in animals and can spread to other susceptible organisms through direct contact or through consumption of animal products), reservoirs (resistant bacteria lives and multiplies in animals and are totally dependent on them for survival) and direct disseminators of resistant bacteria and genes to other animals, humans, and the environment. Animals upon consumption of antibiotics release antibiotic residues and superbugs in the environment in the form of waste or as animal products such as milk, and eggs. The reckless use of unapproved antibiotics in the integrated practice of poultry and aquaculture enhances the chances of spread of antibiotic resistance from one system to another. These residues and resistant bacteria can reach humans through the consumption of improperly cooked animal products and through direct contact with animals. If animal waste and aquaculture waste are used as manure in plant farming, then the superbugs and antibiotic residues get transferred to humans through consumption of vegetables and fruits grown in the farm as well as through consumption of aquaculture products. The below image (Figure 15.5) provides a snapshot of the spread of AMR in the animal sector.

The resistant bacteria can get transferred from animals to humans and the other way around. Studies show that the one risk factor for methicillin resistant *Staphylococcus aureus* (MRSA) infection in humans is direct contact with pet animals like horses, pigs, cattle, dogs and cats that have MRSA strain within them [10, 11]. Similarly, studies have also found that humans are the major source of MRSA in pet animals [11]. Additionally, animal feeds containing antibiotics as well as the resistant bacteria can also promote the dissemination of AMR. For example, the data from a feed surveillance program by Food and Drug Administration (FDA) identified that 21% of the Salmonella isolates retrieved from pet feeds were resistant to at least one antibiotic [12].

15.3.3.1 Risk factors

Misuse of antibiotics The antibiotics, are given to animals mixed with feed or via. injections. These antibiotics are given for therapeutic, prophylactic and metaphylactic measures in animals. The therapeutic approach corresponds to the use of antibiotics when

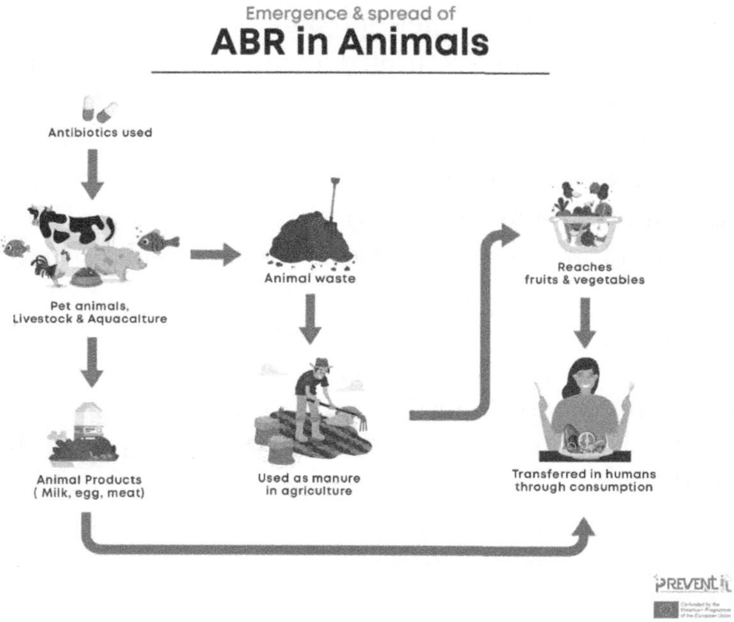

FIGURE 15.5
Spread of antibiotic resistant bacteria and antibiotic residues in animals.

the animals have suffered from disease. The prophylactic approach corresponds towards more prevention, i.e., the disease has not occurred yet. The metaphylactic approach means that few animals are already diseased, and medicine is given to both diseased and non-diseased to avoid spread. The latter two are more problematic because the infectious agent or disease is not known, and the medicines are given without prescription. Further, the use of antibiotics also promotes the growth of the animal that helps farmers to grow their business better.

Increased demand

The increasing population of humankind is directly linked to the increasing demand of food animals and food products. Globally, the number of antimicrobials used in food animals will increase from 63,151 tons in 2010 to 105,596 tons by 2030 - an increase of 67% (Van Boeckel et al. 2015). Let us take an example of India to understand this scenario. It is well known that India's livestock sector accounts for 11.6% of the total livestock population, one of the largest in the world. Additionally, India is the largest producer of milk, third-largest egg producer and the fourth-largest chicken producer after China, Brazil, and USA. In India, the annual fisheries production increased from 0.75 million tonnes in 1950–1951 to 10.79 million tonnes in 2015–2016 [9].

Lack of hygiene

Generally, most of the farmers do not keep the animals in proper hygienic conditions. This situation further aggravates the increasing spread of disease among them which lead to the misuse of antibiotics.

Globalization

As to humans, there is an increase in international trade of animals and/or animal feed. These movements have merged the global space, leading to increased interactions among and between people, animals, animal products and environment.

15.3.4 Plant/ environment domain

Antibiotic use in plants is still a controversial topic with a lot of research gaps to be filled. In plants, antibiotics like streptomycin are used for prophylaxis and therapeutic purposes. Sometimes, antibiotics are mixed with pesticides and applied directly to the crop. The process of AMR development is 100,000 as times rapid in an environment with pesticides and antibiotics together [13]. These residues end up in the plant products, and humans consume those ultimately, thereby acquiring the resistance. Another way of the spread is through soil. Animal waste is used as fertilizer in the agricultural sector. So, drug residues and resistant microbes present in the waste again reach humans via agricultural-plant product consumption. Many municipalities, antibiotic producing pharmaceutical companies, food industries (which use antibiotics as preservatives) and hospitals do not possess appropriate waste management systems. These wastes, if not processed appropriately, contain antibiotic residues and resistant organisms and genes. Such wastes may reach rivers along with rainwater through surface runoff or residues may percolate to groundwater and again reach humans through drinking water or through fruits and vegetables grown in

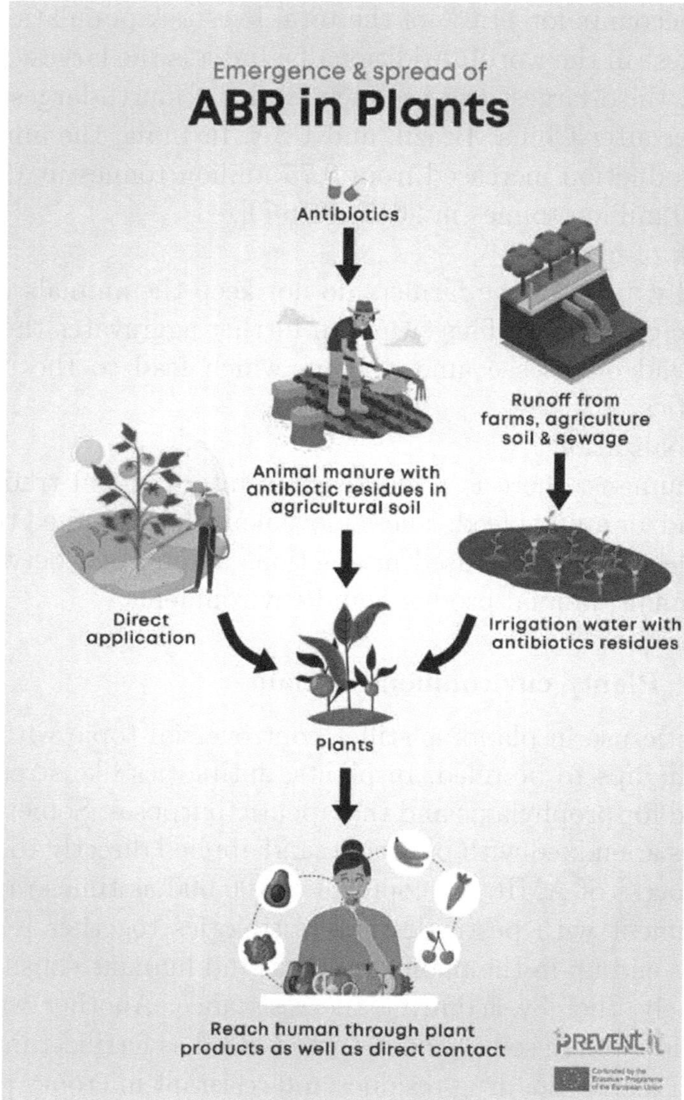

FIGURE 15.6
The emergence and spread of antibiotic resistant bacteria and antibiotic residues in plants.

such water. These pathways are summarized in the image (Figure 15.6) below.

It has been also found that air surrounding waste management systems are polluted with the antibiotic resistant bacteria and may cause drug resistant infection to nearby residents. Microbial aerosols in the air are more diverse than those in water [14]. It is difficult to establish a pivotal relationship between AMR and manure, soil, water, and air due to the complexities of the environment. In addition to above mentioned, heavy metals, detergents and biocides can also co-select for the drug resistant genes [15]. Additionally, encroachment of humans into new habitats have disturbed the environment sustainability, exposing more species to resistant infectious microbes.

15.3.4.1 Risk factors

1. Exposure to antibiotic residues. The antibiotic use in plants is restricted to both therapeutic and prophylactic measures. It has been estimated by FAO that approximately 90% of antibiotics used in animal health care are discharged to environment un-metabolised (antibiotic residues) reaching the plants in the form of manure [16]. For instance, antibiotics like gentamicin in Mexico and Costa Rica, oxonilic acid in Israel and kasugamycin in Japan and other Asian countries are widely used [16].

To conclude, humans and animals are in contact with the resistant bacteria, genes and antibiotic residues through food, water, and soil. Those bacteria in the gut of humans and animals are exposed continuously to antibiotics, creating a vicious cycle leading to exponential increase in the population of resistant organisms in humans and animals. All domains are connected either directly or indirectly with each other, clearly showing the significance of a multisectoral approach to develop interventions targeted against the whole picture rather than the subsections.

15.4 One health and AMR - A few examples

All the above-mentioned factors created the perfect opportunities for the microbes to establish new habitats, spread to various species leading to global presence and to become resistant. Let's take up the example of *E.coli* to learn more. *E.coli* is a common cause of bacterial infections and displays multidrug resistance. The problem of *E.coli* resistance is much more pronounced in developing countries as compared to developed counterparts. The main reservoir of this bacteria is the bowel of humans and animals. Humans carry many acquired *E. coli*, especially the resistant ones. Studies have shown that the travellers from developed countries acquired the drug resistant strains of *E.coli* during their visit to developing countries, and they carried it back to their homeland. The resistant strains persisted in them for more than 6 months [17]. Another example is the spread of antibiotic resistant genes like New Delhi metallo-beta-lactamase enzyme -1 (NDM-1) gene. The first case of NDM-1 gene was reported in 2008 in a Swedish patient, who was previously treated in India. Medical tourism played a key role in spreading the NDM-1 gene from India to the United Kingdom in 2008. Later, in 2010, this gene was identified in the surface and tap water samples in India and in the environmental samples from Bangladesh. Over time, the gene was identified in the arctic soil through the migratory bird wastes [18].

15.5 One health efforts to control AMR

Recognising that AMR is an issue that requires a one health approach, the actions are being undertaken both at global and national levels. Global level - The Food and Agriculture Organization (FAO), World Health Organization (WHO) and World Organization for Animal Health (OIE) along with UNICEF and World Bank developed a policy to reduce the transmission risk of emerging diseases. In the year 2010, FAO, WHO and OIE released a tripartite concept note, wherein they agreed to collaborate, share responsibilities and to coordinate activities to address the health risks at the human-animal-environment juncture [4]. They play an important role in creating and disseminating the right information, networking, and capacity building. The International Ministerial Conference on Animal and Pandemic Influenza at Hanoi in April 2010 recommended broad application of one health approach, which was adopted by the European Union, World Bank and United Nations.

In 2020, One Health Global Leaders Group was created which included representatives from governments, private sector, and civil society [19]. The group aims to provide political leadership to address the issue of drug resistance at global, national, and regional levels. This group brings together the stakeholders from various sectors and advocates for best antibiotic practices across the sectors. The United Nations Environment Programme (UNEP) is coordinates with FAO-WHO-OIE to promote the capacity of the member states on implementing the one health approach and to increase the multi- stakeholder involvement in AMR control. National level - In addition to the above-mentioned organizations, various national and state governments, private sector, civil society, and research institutes are involved in one health activity to prevent and control AMR. Wellcome trust, The AMR Alliance, International Centre for Antimicrobial Resistance Solutions to name a few. The priorities of 'One Health' include surveillance, preventing and managing disease outbreaks, promoting food safety at national and international levels. Using a One Health approach, it helps us to design, implement and monitor programmes, policies, and research on AMR surveillance to provide evidence and also advance multisectoral collaboration to achieve successful outcomes. Successful examples - European experience - Sweden is the first country to ban any non-therapeutic use (including prophylactic and metaphylactic) of antibiotics in livestock. Later in the year 2006, the European Union passed the ban to avoid misuse of antibiotics in all member nations (Khan 2017). US experience - In 2014, the White House issued a national plan to promote the use of antibiotic-free animal products in support of the Food and Drug administration (FDA) [18]. India's experience - Indian government prohibited 'colistin' use in animals as it is the last resort drug for the treatment of *E. coli* infections [18].

15.6 One health approach in AMR- Challenges

There are several challenges to be overcome for successfully implementing the one health concept. Identifying these challenges early on is necessary. These challenges can be understood at three levels [20].

15.6.1 Challenges in developing a one health program

Insufficient funds - The lack of funds seems to be the prime challenge. Most of the one health programmes need a large amount of funding for training the staff for interdisciplinary work, data collection and analysis, hiring the consultants and extending the surveillance program. There are no special budgetary allocations for one health implementation in many countries. The governments across the globe prioritize security and economy over the health of people. Development of infrastructure, laboratory facilities, training of staff also depends on the availability of funds [21, 22].

15.6.1.1 Lack of awareness

Studies have identified that lack of awareness on one health among the practitioners is the next challenge to be addressed. There is a lack of advocacy from the practitioner's side as well on one health. Additionally, there is a lack of commitment from the policy makers because one health requires coordination between various ministries in all aspects which is difficult to achieve [22].

15.6.1.2 Poor governance

Other legislation level challenges in implementing a one health program includes poor governance, poor coordination across sectors, corruption, lack of citizen council, lack of trust and rapid changes in government policies including absence of policies [23]. Additionally, information on the AMR burden in animals and the environment is scanty. Hence, it becomes difficult to convince the policy makers on benefits of one health approach [21, 22]. There is an acute shortage of well-trained one health practitioners to coordinate the projects. Training on One health is not optimal in many developing nations. The academic support for one health project is minimal [21, 22].

15.6.2 Challenges in implementing a one health program

Challenges in implementing a one health program are as follows

15.6.2.1 Lack of surveillance

Though there are established surveillance systems in North America and European regions like European Antimicrobial Resistance Surveillance Network, European Surveillance of Antimicrobial Consumption Network, Transatlantic Taskforce on Antimicrobial Resistance, Danish Integrated Antimicrobial Resistance Monitoring and Research Program, etc., the situation is different in developing countries. An integrated surveillance system on antimicrobial use and resistance across various domains- humans, animals and environment are still absent in many countries. The environment dimension is often missing in most of the existing surveillance systems of developing nations. Hence, the available data are insufficient to compare the resistance patterns among various sectors and to identify critical control points. Considering the existing surveillance systems, most of them lack standardization and harmonization. With the aim to harmonize the surveillance systems between countries, WHO launched Global Antimicrobial Resistance Surveillance system (GLASS) but restricted to humans alone. The quality of surveillance is also affected by logistical issues like lack of resources and workforce. A compromised surveillance system leads to delays in identifying diseases.

15.6.2.2 Less collaborations

One health approach requires collaboration from multiple stakeholders from various sectors. The stakeholders are from varied cultures, language, backgrounds, and have their own perceptions and priorities. It is difficult to promote and sustain such collaborations. Many sectors are still underrepresented in the one health system like environmental scientists, social scientists, economic scientists, anthropological and wildlife scientists. One health is still considered as a "veterinary-driven initiative." Most of the one health practitioners do not include community actors. Hence, they fail to address context specific issues.

15.6.3 Challenges in monitoring and evaluating a one health program

There is a dearth of evaluation studies because most of the one health programs require a long time of monitoring. Similarly, there is an absence of guidelines, qualitative and quantitative indicators for measuring the outcomes in a one health program. Computing the effectiveness and efficiency of a one health intervention becomes challenging in such a situation. This could, in turn, hinder political interest and support.

15.7 Conclusion

The omnipresence of antibiotic resistant bacteria and its spread among humans, animals, plants and environment makes it impossible to treat AMR with a uni-sectoral approach. The One Health approach connecting all the aspects seems the only feasible solution. Of course, governments, public and private sectors have joined hands to achieve this goal, however it demands huge infrastructure, in depth understanding and collaborations. The understanding of 'how much' the problem is will be the first step followed by 'how to stop' the spread among these sectors. Hopefully, the personnel involved will understand the need of the hour and will contribute successfully to lowering the incidence of AMR.

Bibliography

[1] Jim O'Neill. Tackling drug-resistant infections globally: final report and recommendations. 2016.

[2] AVMA. One health- what is one health?, 2022.

[3] World Health Organization et al. Tripartite and unep support ohhlep's definition of "one health". 2021, 2021.

[4] FAOOIE WHO. The fao-oie-who collaboration sharing responsibilities and coordinating global activities to address health risks at the animal-human-ecosystems interfaces. In *International Ministerial Conference on Animal and Pandemic Influenza. Hanoi*, 2010.

[5] Food, Drug Administration, et al. Antimicrobials sold or distributed for use in food-producing animals. *US Food and Drug Administration: Silver Spring, MD, USA*, 2015.

[6] Patricia S McManus, Virginia O Stockwell, George W Sundin, and Alan L Jones. Antibiotic use in plant agriculture. *Annual review of phytopathology*, 40(1):443–465, 2002.

[7] Philip Taylor and Robert Reeder. Antibiotic use on crops in low and middle-income countries based on recommendations made by agricultural advisors. *CABI Agriculture and Bioscience*, 1(1):1–14, 2020.

[8] Bonnie M Marshall and Stuart B Levy. Food animals and antimicrobials: impacts on human health. *Clinical microbiology reviews*, 24(4):718–733, 2011.

[9] Delia Grace. Review of evidence on antimicrobial resistance and animal agriculture in developing countries. 2015.

[10] Inge Van Loo, Xander Huijsdens, Edine Tiemersma, Albert De Neeling, Nienke van de Sande-Bruinsma, Desiree Beaujean, Andreas Voss, and Jan Kluytmans. Emergence of methicillin-resistant staphylococcus aureus of animal origin in humans. *Emerging infectious diseases*, 13(12):1834, 2007.

[11] J Scott Weese and Engeline van Duijkeren. Methicillin-resistant staphylococcus aureus and staphylococcus pseudintermedius in veterinary medicine. *Veterinary microbiology*, 140(3-4):418–429, 2010.

[12] Yi-Yun Liu, Yang Wang, Timothy R Walsh, Ling-Xian Yi, Rong Zhang, James Spencer, Yohei Doi, Guobao Tian, Baolei Dong, Xianhui Huang, et al. Emergence of plasmid-mediated colistin resistance mechanism mcr-1 in animals and human beings in china: a microbiological and molecular biological study. *The Lancet infectious diseases*, 16(2):161–168, 2016.

[13] Brigitta Kurenbach, Amy M Hill, William Godsoe, Sophie van Hamelsveld, and Jack A Heinemann. Agrichemicals and antibiotics in combination increase antibiotic resistance evolution. *PeerJ*, 6:e5801, 2018.

[14] Jeadran N Malagón-Rojas, Eliana L Parra Barrera, and Luisa Lagos. From environment to clinic: the role of pesticides in antimicrobial resistance. *Revista Panamericana de Salud Pública*, 44, 2020.

[15] Bjorn JA Berendsen, Robin S Wegh, Joost Memelink, Tina Zuidema, and Linda AM Stolker. The analysis of animal faeces as a tool to monitor antibiotic usage. *Talanta*, 132:258–268, 2015.

[16] FAO. Antimicrobial resistance in food and agriculture. Report, 2017.

[17] John S Mackenzie, Martyn Jeggo, Peter Daszak, and Juergen A Richt. *One Health: The human-animal-environment interfaces in emerging infectious diseases*, volume 366. Springer, 2013.

[18] Asad U Khan, Lubna Maryam, and Raffaele Zarrilli. Structure, genetics and worldwide spread of new delhi metallo-β-lactamase (ndm): a threat to public health. *BMC microbiology*, 17(1):1–12, 2017.

[19] Wellcome Trust. The global response to amr momentum, success, and critical gaps, 2020.

[20] Carolina dos S Ribeiro, Linda HM van de Burgwal, and Barbara J Regeer. Overcoming challenges for designing and implementing the one health approach: A systematic review of the literature. *One Health*, 7:100085, 2019.

[21] Samer Abuzerr, Kate Zinszer, and Abraham Assan. Implementation challenges of an integrated one health surveillance system in humanitarian settings: A qualitative study in palestine. *SAGE open medicine*, 9:20503121211043038, 2021.

[22] Krishna Prasad Acharya, Surendra Karki, Kshitiz Shrestha, and Krishna Kaphle. One health approach in nepal: Scope, opportunities and challenges. *One Health*, 8:100101, 2019.

[23] Esther Buregyeya, Edwinah Atusingwize, Peninah Nsamba, David Musoke, Irene Naigaga, John David Kabasa, Hellen Amuguni, and William Bazeyo. Operationalizing the one health approach in uganda: Challenges and opportunities. *Journal of Epidemiology and Global Health*, 10(4):250, 2020.

Index

For Product Safety Concerns and Information please contact our EU representative GPSR@taylorandfrancis.com Taylor & Francis Verlag GmbH, Kaufingerstraße 24, 80331 München, Germany